Lecture Notes in Computer Science 6519

Commenced Publication in 1973
Founding and Former Series Editors:
Gerhard Goos, Juris Hartmanis, and Jan van Leeuwen

Wim van Dam   Vivien M. Kendon
Simone Severini (Eds.)

Theory of Quantum Computation, Communication, and Cryptography

5th Conference, TQC 2010
Leeds, UK, April 13-15, 2010
Revised Selected Papers

 Springer

Volume Editors

Wim van Dam
University of California
Department of Computer Science
Santa Barbara, CA 93106-5110, USA
E-mail: vandam@cs.ucsb.edu

Vivien M. Kendon
University of Leeds
School of Physics and Astronomy
Leeds, LS2 9JT, UK
E-mail: v.kendon@leeds.ac.uk

Simone Severini
University College London
Department of Physics and Astronomy
London, WC1E 6BT, UK
E-mail: simoseve@gmail.com

Library of Congress Control Number: 2010941752

CR Subject Classification (1998): F, D, C.2, G.1-2, E.3, J.2

LNCS Sublibrary: SL 1 – Theoretical Computer Science and General Issues

ISSN	0302-9743
ISBN-10	3-642-18072-8 Springer Berlin Heidelberg New York
ISBN-13	978-3-642-18072-9 Springer Berlin Heidelberg New York

springer.com

© Springer-Verlag Berlin Heidelberg 2011
Printed in Germany

Typesetting: Camera-ready by author, data conversion by Scientific Publishing Services, Chennai, India
Printed on acid-free paper   06/3180

Preface

The Conference on Theory of Quantum Computation, Communication, and Cryptography (TQC) is an annual meeting on theoretical aspects of quantum information processing. The goal of the conference is to foster developments in this rapidly growing, interdisciplinary field by providing a forum for the presentation and discussion of original research.

The fifth iteration of TQC was held during April 13–15, 2010, at the University of Leeds, United Kingdom. It included invited talks, contributed talks, and a poster session, as well as a rump session consisting of short talks on recent developments. Authors of selected contributed talks were invited to submit a paper to these proceedings.

TQC 2010 would not have been possible without the contributions of numerous individuals and organizations, and we sincerely thank them for their support.

In putting together the scientific program, we were very grateful for the hard work and advice of the Program Committee, listed herein. The logistics of the conference were expertly managed by the Organizing Committee, also listed herein, and we thank them for their efforts to make the conference a success.

We would like to thank the invited speakers, Frédéric Magniez, Kae Nemoto, Frank Verstraete, Ronald de Wolf, and Anton Zeilinger, for their contributions to the program.

We would like to thank the members of the Conference Series Steering Committee, Yasuhito Kawano, Michele Mosca, and Vlatko Vedral, for their important advice.

TQC 2010 was made possible by financial support from the British Computer Society, the Heilbronn Institute, the Quantum Information, Quantum Optics and Quantum Control Group of the Institute of Physics, the School of Mathematics of the University of Leeds, the School of Physics and Astronomy of the University of Leeds, the London Mathematical Society, the Sandia National Laboratories, the Institute for Quantum Computing at the University of Waterloo, and the Worldwide Universities Network, Leeds; we thank these organizations for their important contributions.

Finally, we would like to thank Springer for publishing the proceedings of TQC in the *Lecture Notes in Computer Science* series.

October 2010

Wim van Dam
Vivien Kendon
Simone Severini

Organization

Program Committee

Wim van Dam	University of California, Santa Barbara (Chair), USA
Simone Severini	University College London (Co-chair), UK
Dagmar Bruß	Heinrich Heine University, Germany
Andrew Childs	University of Waterloo, Canada
Matthias Christandl	Ludwig Maximilians University, Germany
Nilanjana Datta	University of Cambridge, UK
Aram Harrow	University of Bristol, UK
Peter Høyer	University of Calgary, Canada
Rahul Jain	National University of Singapore
Elham Kashefi	University of Edinburgh, UK
Debbie Leung	University of Waterloo, Canada
Hoi-Kwong Lo	University of Toronto, Canada
Juan Pablo Paz	University of Buenos Aires, Argentina
Francesco Petruccione	University of KwaZulu-Natal, South Africa
David Poulin	Université de Sherbrooke, Canada
Martin Rötteler	NEC, Princeton, USA
Miklos Santha	Université Paris Sud, France
Seiichiro Tani	NTT, Tokyo, Japan
Jean-Pierre Tillich	INRIA, Rocquencourt, France
Pawel Wocjan	University of Central Florida, USA

Organizing Committee

Vivien Kendon	University of Leeds (Chair), UK
Martin Aulbach	University of Leeds, UK
Dave Bacon	University of Washington, USA
Stephen Bartlett	University of Sydney, Australia
Katie Barr	University of Leeds, UK
Stephen Brierley	University of York, UK
Katherine Brown	University of Leeds, UK
Barry Cooper	University of Leeds, UK
Peter Crompton	University of Leeds, UK
Vladimir V. Kisil	University of Leeds, UK
Neil Lovett	University of Leeds, UK
Stefano Pirandola	University of York, UK
Mike Stannett	University of Sheffield, UK
Rob Wagner	University of Leeds, UK

Table of Contents

Asymptotically Optimal Discrimination between Pure Quantum States

Michael Nussbaum[1],[*] and Arleta Szkoła[2]

[1] Department of Mathematics, Cornell University, Ithaca NY, USA
[2] Max Planck Institute for Mathematics in the Sciences, Leipzig, Germany

Abstract. We consider the decision problem between a finite number of states of a finite quantum system, when an arbitrarily large number of copies of the system is available for measurements. We provide an upper bound on the exponential rate of decay of the averaged probability of rejecting the true state. It represents a generalized quantum Chernoff distance of a finite set of states. As our main result we prove that the bound is sharp in the case of pure states.

Keywords: multiple quantum state discrimination, generalized quantum Chernoff distance, quantum hypothesis testing, error exponents.

1 Introduction

In various branches of quantum theory such as quantum information processing, quantum communication theory or quantum statistics one of the basic problems is to determine the state of a given quantum system. In the simplest case there is a *finite* set of states specifying the possible preparation of the quantum system. In the Bayesian approach of quantum statistics, the likelihood of the different states is determined by an a priori probability distribution. One makes a decision in favor of one of the states following a specified rule based on the outcomes of a generalized measurement -called a quantum test. In the binary case optimal tests, i.e. tests minimizing the averaged probability of rejecting the true state, are known to be given by Holevo-Helstrom projections [5], [4]. These generalize the classical likelihood ratio tests. Here we consider the scenario where there is an arbitrarily large finite number n of copies of the quantum system available for performing a measurement. The corresponding state is then described by an n-fold tensor product of one of the associated density operators. There are two main goals: firstly, to construct a sequence of quantum tests in n which maximize the asymptotic (exponential) rate of decay of the averaged probability of rejecting the true state. The second goal is to determine the corresponding optimal error exponent. It has been shown that in the binary case asymptotically optimal quantum tests, thus in particular the Holevo-Helstrom tests, achieve an exponential rate of decay which is equal to the quantum Chernoff bound, cf. [8], [1] and [2]. Surprisingly, the corresponding questions in the case of $r > 2$ states

---

[*] Supported in part by NSF grant DMS-08-05632.

have not yet received a final answer, despite a number of efforts and numerous strong results obtained in relation to multiple quantum state discrimination, see [11], [7], [3], [10] and references therein.

We define a generalized quantum Chernoff distance of a finite set of states as the minimum of the binary quantum Chernoff distances over all possible pairs of different states. The binary quantum Chernoff distance has been introduced in the context of binary quantum hypothesis testing in [8]. Relying on [8] we prove that the generalized quantum Chernoff distance specifies a bound on the achievable asymptotic error exponents in multiple quantum state discrimination. This is in line with results obtained in the context of classical multiple hypothesis testing, cf. [9]. As our main result we prove that in the special case of pure quantum states this bound, indeed, is achievable and hence specifies the optimal asymptotic error exponent. The corresponding asymptotically optimal quantum tests rely on a Gram-Schmidt orthonormalization procedure of the associated state vectors. Similar quantum tests were already considered by Holevo in [6] in the context of quantum minimal error decision problems. However, the question of the corresponding asymptotic error exponent is not addressed in [6].

2   Notations and the Main Results

Let S be a finite quantum system and $\mathcal{H}$ be the associated complex Hilbert space with $\dim \mathcal{H} = d < \infty$. Further denote by $\mathcal{A}$ the algebra of observables of S, i.e. $\mathcal{A}$ is the algebra of linear operators on $\mathcal{H}$. For each $n \in \mathbb{N}$ denote by $\mathcal{A}^{(n)}$ the algebra of linear operators on the n-fold tensor product Hilbert space $\mathcal{H}^{\otimes n}$. It represents the algebra of observables of a compound quantum system S_n with its n unit systems being of the same type S.

For each $n \in \mathbb{N}$ the set of density operators in $\mathcal{A}^{(n)}$ corresponds one-to-one to the state space $\mathcal{S}(\mathcal{A}^{(n)})$ of $\mathcal{A}^{(n)}$. Recall that a density operator is defined to be a self-adjoint, positive linear operator of trace 1.

Let $r \in \mathbb{N}$ and Σ be a set of density operators $\rho_i \in \mathcal{S}(\mathcal{A})$, $i = 1, \ldots, r$, representing the possible states of the quantum system S. Assume that for each $n \in \mathbb{N}$ there is a compound quantum system S_n being an n-fold copy of S. This means, in particular, that the corresponding quantum state is in $\Sigma^{\otimes n} := \{\rho_i^{\otimes n}\}_{i=1}^r$, i.e. it is uniquely determined by the index $i \in \{1, \ldots, r\}$.

Further, let $E^{(n)} = \{E_i^{(n)}\}_{i=1}^r$ be a positive operator valued measure (POVM) in $\mathcal{A}^{(n)}$, i.e. each $E_i^{(n)}$, $i = 1, \ldots, r$, is a self-adjoint element of $\mathcal{A}^{(n)}$ with $E_i^{(n)} \geq 0$ and $\sum_{i=1}^r E_i^{(n)} = \mathbf{1}$. The POVMs $E^{(n)}$ describe quantum tests for discrimination between the r states from $\Sigma^{\otimes n}$, or simply *quantum tests for* $\Sigma^{\otimes n}$, by identifying the measurement outcome corresponding to $E_i^{(n)}$, $i = 1, \ldots, r$, with the density operator $\rho_i^{\otimes n}$, respectively. If ρ_i happens to describe the true state of S, and correspondingly $\rho_i^{\otimes n}$ determines the state of S_n, then the associated *individual success probability* is given by

$$\mathrm{Succ}_i(E^{(n)}) := \mathrm{tr}\,[\rho_i^{\otimes n} E_i^{(n)}]\,.$$

The *individual error probability* refers to the situation when the density operator ρ_i is discarded as possible preparation of S; it is given by the formula

$$\mathrm{Err}_i(E^{(n)}) := \mathrm{tr}\,[\rho_i^{\otimes n}(1 - E_i^{(n)})]\ .$$

Assuming $0 < p_i < 1$, $i = 1, \ldots, r$, with $\sum_{i=1}^r p_i = 1$ to be the a priori distribution of the r quantum states from Σ the *averaged error probability* is defined by

$$\mathrm{Err}(E^{(n)}) = \sum_{i=1}^r p_i \mathrm{tr}\,[\rho_i^{\otimes n}(1 - E_i^{(n)})]\ .$$

If the limit $\lim_{n\to\infty} -\frac{1}{n}\log \mathrm{Err}(E^{(n)})$ exists, we refer to it as the *asymptotic error exponent*. Otherwise we have to consider the corresponding lim sup and lim inf expressions.

For two density operators ρ_1 and ρ_2 the *quantum Chernoff distance* is defined by

$$\xi_{QCB}(\rho_1, \rho_2) := -\log \inf_{0 \le s \le 1} \mathrm{tr}\,[\rho_1^{1-s}\rho_2^s]\ . \tag{1}$$

It specifies the optimal achievable asymptotic error exponent in discriminating between ρ_1 and ρ_2, compare [8], [1], [2]. Quantum tests with minimal averaged error probability for a pair of different density operators ρ_1 and ρ_2 on the same Hilbert space $\mathcal{H}$ are well-known to be given by the respective *Holevo-Helstrom projectors*

$$\Pi_1 := \mathrm{supp}\,(\rho_1 - \rho_2)_+, \quad \Pi_2 := \mathrm{supp}\,(\rho_2 - \rho_1)_+ = 1 - \Pi_1\ .$$

Here supp a denotes the support projector of a self-adjoint operator a, while a_+ means its positive part, i.e. $a_+ = (|a| + a)/2$ for $|a| := (a^*a)^{1/2}$, see [5], [4]. As mentioned in the introduction, the Holevo-Helstrom projectors generalize the likelihood ratio tests for two probability distributions. This can be verified by letting ρ_1 and ρ_2 be two commuting density matrices, cf. [8].

For a set $\Sigma = \{\rho_i\}_{i=1}^r$ of density operators on $\mathcal{H}$, where $r > 2$, we introduce the *generalized quantum Chernoff distance*

$$\xi_{QCB}(\Sigma) := \min\{\xi_{QCB}(\rho_i, \rho_j) : \ 1 \le i < j \le r\}\ . \tag{2}$$

This is in full analogy to the definition of the generalized Chernoff distance in classical multiple hypothesis testing, where the density operators are replaced by probability distributions on a finite sample space, cf. [9].

Our first theorem is an implication of Theorem 2.2 in [8].

Theorem 1. *Let $r \in N$ and $\Sigma = \{\rho_i\}_{i=1}^r$ be a set of pairwise different density operators on $\mathcal{H}$ with corresponding a priori probability distribution $\{p_i\}_{i=1}^r$. For any sequence $E^{(n)}$, $n \in N$, of quantum tests for $\Sigma^{\otimes n}$, respectively, it holds*

$$\limsup_{n\to\infty} -\frac{1}{n}\log \mathit{Err}(E^{(n)}) \le \xi_{QCB}(\Sigma)\ ,$$

where $\xi_{QCB}(\Sigma)$ is the generalized quantum Chernoff distance defined by (2).

It turns out that the generalized quantum Chernoff distance is achievable as an asymptotic error exponent in the case of pure states. This is the statement of our main theorem below.

Theorem 2. *Let $r \in \mathbb{N}$ and $\Sigma = \{\rho_i\}_{i=1}^r$ be a set of pairwise different pure states of a quantum system S. Then there exists a sequence $\{E^{(n)}\}_{n \in \mathbb{N}}$ of quantum tests for $\Sigma^{\otimes n}$, respectively, with*

$$\lim_{n \to \infty} -\frac{1}{n} \log Err(E^{(n)}) = \xi_{QCB}(\Sigma) \ ,$$

i.e. the generalized quantum Chernoff distance is an achievable asymptotic error exponent in multiple pure state discrimination.

3   Generalized Quantum Chernoff Bound in Multiple Quantum State Discrimination

In this section we give a proof of Theorem 1 stating that the generalized quantum Chernoff distance specifies a bound on the asymptotically achievable error exponents in multiple quantum state discrimination. It relies on its binary version presented in Theorem 2.2 in [8].

Proof (Theorem 1). Fix any two indicies $1 \le i < j \le r$. For $n \in \mathbb{N}$ let $A^{(n)}, B^{(n)} \in \mathcal{A}^{(n)}$ be two positive operators such that $A^{(n)} + B^{(n)} = \mathbf{1} - E_i^{(n)} - E_j^{(n)}$. Then the positive operators $\tilde{E}_i^{(n)} := E_i^{(n)} + A^{(n)}$ and $\tilde{E}_j^{(n)} := E_j^{(n)} + B^{(n)}$ represent a POVM $\tilde{E}^{(n)}$ in $\mathcal{A}^{(n)}$, which we consider a quantum test for the pair $\{\rho_i^{\otimes n}, \rho_j^{\otimes n}\}$. For the individual error probabilities of the modified quantum test $\tilde{E}^{(n)}$ we obtain the upper bounds

$$\mathrm{Err}_i(\tilde{E}^{(n)}) = \mathrm{tr}\,[\rho_i^{\otimes n}(\mathbf{1} - \tilde{E}_i^{(n)})] \le \mathrm{tr}\,[\rho_i^{\otimes n}(\mathbf{1} - E_i^{(n)})] = \mathrm{Err}_i(E^{(n)}) \ ,$$

and similarily $\mathrm{Err}_j(\tilde{E}^{(n)}) \le \mathrm{Err}_j(E^{(n)})$. It follows a lower bound on the average error probability with respect to the original tests $\{E_i^{(n)}\}_{i=1}^r$:

$$\mathrm{Err}(E^{(n)}) = \sum_{k=1}^r p_k \mathrm{Err}_k(E^{(n)}) \ge \left(p_i \mathrm{Err}_i(E^{(n)}) + p_j \mathrm{Err}_j(E^{(n)}) \right)$$

$$\ge \left(p_i \mathrm{Err}_i(\tilde{E}^{(n)}) + p_j \mathrm{Err}_j(\tilde{E}^{(n)}) \right)$$

$$\ge p_{\min} \left(\mathrm{Err}_i(\tilde{E}^{(n)}) + \mathrm{Err}_j(\tilde{E}^{(n)}) \right) \ ,$$

where $p_{\min} := \min\{p_i : 1 \le i \le r\}$. The above bound implies

$$\limsup_{n \to \infty} -\frac{1}{n} \log \mathrm{Err}(E^{(n)}) \le \limsup_{n \to \infty} -\frac{1}{n} \log p_{\min}$$

$$+ \limsup_{n \to \infty} -\frac{1}{n} \log \left(\mathrm{Err}_i(\tilde{E}^{(n)}) + \mathrm{Err}_j(\tilde{E}^{(n)}) \right)$$

$$= \limsup_{n \to \infty} -\frac{1}{n} \log \frac{1}{2} \left(\mathrm{Err}_i(\tilde{E}^{(n)}) + \mathrm{Err}_j(\tilde{E}^{(n)}) \right)$$

$$\le \xi_{QCB}(\rho_i, \rho_j) \ .$$

Here the last inequality is by Theorem 2.2 in [8], which represents the statement of our Theorem 1 in its binary version corresponding to the special case $r = 2$. Since the pair of indicies (i, j) was choosen arbitrary, the statement of the theorem follows. □

4 Asymptotically Optimal Pure State Discrimination

In this section we provide a constructive proof for Theorem 2. Roughly speaking, our quantum tests, which can be shown to achieve an asymptotic error exponent equal to the generalized quantum Chernoff distance of Σ, are obtained from a Gram-Schmidt orthonormalization procedure of the unit vectors associated to the pure states in Σ.

Proof (Theorem 2). Observe that in view of Theorem 1 it is sufficient to construct quantum tests for which we can verify

$$\liminf_{n \to \infty} -\frac{1}{n} \log \mathrm{Err}(E^{(n)}) \geq \xi_{QCB}(\Sigma) \ .$$

For each $1 \leq i \leq r$ let v_i be a unit vector in $\mathcal{H}$ such that $|v_i\rangle\langle v_i| = \rho_i$.

1. We assume that the set $V(\Sigma) := \{v_i\}_{i=1}^r$ is linearly independent and start with the case $n = 1$, where no tensor products are included. We define for each $k = 1, \ldots, r$, a $(d \times k)$ matrix Ψ_k

$$\Psi_k := (v_1, \ldots, v_k) \ , \tag{3}$$

i.e. the columns of Ψ_k are equal to the state vectors v_i, $1 \leq i \leq k$. We refer to the $(k \times k)$-matrix

$$\Psi_k^* \Psi_k =: \Gamma_k$$

as a Gram matrix of $\{v_1, \ldots, v_k\}$. By the assumption of linear independence of the set $V(\Sigma)$ for each $k \in \{1, \ldots, r\}$ the operator

$$P_k := \Psi_k (\Psi_k^* \Psi_k)^{-1} \Psi_k^* = \Psi_k \Gamma_k^{-1} \Psi_k^* \ ,$$

represents an orthogonal projector onto a k-dimensional subspace of $\mathcal{H}$, which is spanned by the k state vectors $v_1, \ldots, v_k$. Further, we set $P_0 = 0$ and define for $1 \leq k \leq r$

$$E_k := P_k - P_{k-1} \ .$$

The E_k represent one-dimensional orthogonal projectors, which are mutually orthogonal. With $e_k := \frac{1}{\|E_k v_k\|} E_k v_k$ we can write $E_k = |e_k\rangle\langle e_k|$, and the set $\{e_k\}_{k=1}^r$ represents a Gram-Schmidt orthonormalization of the linearly independent set $V(\Sigma)$ of unit vectors v_k, $k = 1, \ldots, r$.

Observe that by construction $\sum_{i=1}^{r} E_i \leq 1$. If $E_0 := 1 - \sum_{i=1}^{r} E_i \neq 0$, we redefine E_1 to be $E_1 + E_0$, such that $\sum_{i=1}^{r} E_i = 1$ is satisfied. By identifying E_i, $i = 1, \ldots, r$, with ρ_i, respectively, we obtain a quantum test $E^{(1)} = \{E_i\}_{i=1}^{r}$ for Σ.

For $1 \leq i \leq r$ the corresponding individual success probability reads

$$\mathrm{Succ}_i(E^{(1)}) = \mathrm{tr}\,[\rho_i E_i] = \mathrm{tr}\,[|v_i\rangle\langle v_i|E_i] = \langle v_i|P_i - P_{i-1}|v_i\rangle \ . \tag{4}$$

Since the P_i's are constructed as orthogonal projectors onto $\mathrm{span}\{v_1, \ldots, v_i\}$ it holds $|v_i\rangle\langle v_i| \leq P_i$ and as a consequence $\langle v_i|P_i|v_i\rangle = 1$. Then from the relation $\mathrm{Err}_i(E^{(1)}) = 1 - \mathrm{Succ}_i(E^{(1)})$ we obtain

$$\begin{aligned}
\mathrm{Err}_i(E^{(1)}) &= \langle v_i|P_{i-1}|v_i\rangle \\
&= \langle v_i|\Psi_i(\Gamma_{i-1})^{-1}\Psi_{i-1}^*|v_i\rangle \\
&\leq \frac{1}{\lambda_{\min}(\Gamma_{i-1})}\langle v_i|\Psi_{i-1}\Psi_{i-1}^*|v_i\rangle \\
&= \frac{1}{\lambda_{\min}(\Gamma_{i-1})}\|\Psi_{i-1}^*v_i\|^2 \ ,
\end{aligned} \tag{5}$$

where $\lambda_{\min}(\cdot)$ denotes the minimal eigenvalue of a self-adjoint matrix. By definition (3) of Ψ_i we have

$$\|\Psi_{i-1}^*v_i\|^2 = \sum_{j=1}^{i-1}|\langle v_j|v_i\rangle|^2, \qquad i = 2, \ldots, r \ . \tag{6}$$

Inserting expression (6) into (5) we obtain the upper bound

$$\mathrm{Err}_i(E^{(1)}) \leq \sum_{j=1}^{i-1}\frac{|\langle v_j|v_i\rangle|^2}{\lambda_{\min}(\Gamma_{i-1})} \ . \tag{7}$$

Recall that the density operators ρ_i, $i = 1, \ldots, r$, are expected to appear with probability p_i, respectively. Then the averaged error probability can be estimated from above as follows

$$\begin{aligned}
\mathrm{Err}(E^{(1)}) = \sum_{i=1}^{r} p_i \mathrm{Err}_i(E^{(1)}) &\leq \sum_{i=1}^{r}\mathrm{Err}_i(E^{(1)}) \\
&\leq \sum_{i=2}^{r}\sum_{j=1}^{i-1}\frac{|\langle v_j|v_i\rangle|^2}{\lambda_{\min}(\Gamma_{i-1})} \ ,
\end{aligned} \tag{8}$$

where in the second line we have applied (7).

2. Let $n > 1$. Notice that still assuming that $V(\Sigma)$ is a set of r linearly independent unit vectors, the same remains true for $V(\Sigma^{\otimes n})$ consisting of the n-fold tensor product state vectors $v_i^{\otimes n}$, $i = 1, \ldots, r$. Hence we can adopt the construction of the quantum test $E^{(1)}$ for Σ as it stands for the tensor product

case. In particular, we define $\Psi_{j,n}$, $1 \leq j \leq r$, analogously to (3) as the $(d^n \times j)$-matrix

$$\Psi_{j,n} := \left(v_1^{\otimes n}, \ldots, v_j^{\otimes n} \right) ,$$

respectively. Then the corresponding averaged error probability $\mathrm{Err}\,(E^{(n)})$ can be upper bounded similarily to (8):

$$\mathrm{Err}\,(E^{(n)}) \leq \sum_{i=2}^{r} \sum_{j=1}^{i-1} \frac{|\langle v_j^{\otimes n}|v_i^{\otimes n}\rangle|^2}{\lambda_{\min}(\Gamma_{i-1,n})} = \sum_{i=2}^{r} \sum_{j=1}^{i-1} \frac{\left(|\langle v_j|v_i\rangle|^2\right)^n}{\lambda_{\min}(\Gamma_{i-1,n})} ,$$

where $\Gamma_{i-1,n} := \Psi_{i-1,n}^* \Psi_{i-1,n}$.

Observe that each Gram matrix $\Gamma_{j,n} = \Psi_{j,n}^* \Psi_{j,n}$, $j = 1, \ldots, r$, is a square matrix of fixed dimension j, respectively. Further, note that the diagonal entries $\gamma_{kk}^{(j,n)}$, $k = 1, \ldots, j$ of $\Gamma_{j,n}$ are given by $\langle v_k^{\otimes n}|v_k^{\otimes n}\rangle$, respectively, and hence are all equal to 1. Since for $k \neq l$ it holds $|\langle v_k|v_l\rangle| < 1$, the off-diagonal entries $\gamma_{k,l}^{(j,n)} = \langle v_k^{\otimes n}|v_l^{\otimes n}\rangle = \langle v_k|v_l\rangle^n$ tend to 0 as n goes to infinity. It follows for every $1 \leq j \leq r$

$$\Gamma_{j,n} \to I_j \quad \text{as } n \to \infty ,$$

where I_j denotes the identity matrix of dimension j. By continuity of the minimal eigenvalue this implies

$$\lambda_{\min}(\Gamma_{j,n}) \to 1 \quad \text{as } n \to \infty .$$

We conclude

$$\mathrm{Err}\,(E^{(n)}) \leq \sum_{i=2}^{r} \sum_{j=1}^{i-1} \left(|\langle v_j|v_i\rangle|^2\right)^n (1 + o(1)) .$$

As n tends to infinity the largest term dominates. As a consequence we have

$$\frac{1}{n} \log \mathrm{Err}\,(E^{(n)}) \leq \max\{\log |\langle v_j|v_i\rangle|^2 : 1 \leq j < i \leq r\} + o(1)$$

$$= -\min\{\xi_{QCB}(\rho_i, \rho_j), \, 1 \leq j < i \leq r\} + o(1)$$

$$= -\xi_{QCB}(\Sigma) + o(1) , \tag{9}$$

where in the second line we have used the fact that in the case of two different pure states on $\mathcal{H}$, say $\rho = |v\rangle\langle v|$ and $\sigma = |w\rangle\langle w|$, the corresponding (binary) quantum Chernoff distance $\xi_{QCB}(\rho, \sigma)$ takes the simple form $-\log|\langle v|w\rangle|^2$, cf. [8]. The last identity is by definition (2) of the generalized quantum Chernoff distance. The proof is complete under the assumption of linear independence of the set of eigenvectors of Σ.

3. Finally, notice that even if $V(\Sigma)$ is not linearly independent, the set $V(\Sigma^{\otimes N})$ consisting of N-fold tensor product vectors becomes linearly independent for N large enough. Then, for every $n \geq N$ we can adopt the construction of quantum tests $E^{(n)}$ for $\Sigma^{\otimes n}$ as presented in parts 1 and 2 of the proof, and the asymptotic relation (9) remains valid. □

Acknowledgments. The work of M. N. has been supported in part by NSF Grant DMS-08-05632. A. S. wishes to thank the research groups of Prof. Jost and Nihat Ay at the MPI MiS for their interest and helpful discussions.

References

1. Audenaert, K.M.R., Casamiglia, J., Munoz-Tapia, R., Bagan, E., Masanes, L.l., Acin, A., Verstraete, F.: Discriminating States: The Quantum Chernoff Bound. Phys. Rev. Lett. 98, 160501 (2007)
2. Audenaert, K.M.R., Nussbaum, M., Szkoła, A., Verstraete, F.: Asymptotic Error Rates in Quantum Hypothesis Testing. Commun. Math. Phys. 279, 251–283 (2008)
3. Barett, S., Croke, S.: On the conditions for discrimination between quantum states with minimum error. J. Phys. A: Math. Theor. 42 (2009)
4. Helstrom, C.W.: Quantum Detection and Estimation Theory. Academic Press, New York (1976)
5. Holevo, Λ.: Investigations in the general theory of statistical decisions. Trudy Mat. Inst. Steklov 124 (in Russian) (English translation in Proc. Steklov Inst. of Math. 3. Amer. Math. Soc., Providence) (1978)
6. Kholevo, A.: On asymptotically optimal hypothesis testing in quantum statistics. Theor. Probab. Appl. 23, 411–415 (1978)
7. König, R., Renner, R., Schaffner, C.: The operational meaning of min- and max-entropy. IEEE Trans. Inf. Th. 55(9) (2009)
8. Nussbaum, M., Szkoła, A.: The Chernoff lower bound for symmetric quantum hypothesis testing. Ann. Stat. 37(2), 1040–1057 (2009)
9. Salikhov, N.P.: On one generalisation of Chernov's distance. Theory Probab. Appl. 43(2), 239–255 (1997)
10. Tyson, J.: Two-sided estimates of minimum-error distinguishability of mixed quantum states via generalized Holevo-Curlander bounds. J. Math. Phys. 50, 32106 (2009)
11. Yuen, H.P., Kennedy, R.S., Lax, M.: Optimum testing of Multiple Hypotheses in Quantum Detection Theory. IEEE Trans. Inform. Thoery IT-21(2), 125–134 (1975)

On Quantum Estimation, Quantum Cloning and Finite Quantum de Finetti Theorems

Giulio Chiribella

Perimeter Institute for Theoretical Physics, 31 Caroline Street North, Waterloo,
Ontario N2L 2Y5, Canada

Abstract. This paper presents a series of results on the interplay between quantum estimation, cloning and finite de Finetti theorems. First, we consider the measure-and-prepare channel that uses optimal estimation to convert M copies into k approximate copies of an unknown pure state and we show that this channel is equal to a random loss of all but s particles followed by cloning from s to k copies. When the number k of output copies is large with respect to the number M of input copies the measure-and-prepare channel converges in diamond norm to the optimal universal cloning. In the opposite case, when M is large compared to k, the estimation becomes almost perfect and the measure-and-prepare channel converges in diamond norm to the partial trace over all but k systems. This result is then used to derive de Finetti-type results for quantum states and for symmetric broadcast channels, that is, channels that distribute quantum information to many receivers in a permutationally invariant fashion. Applications of the finite de Finetti theorem for symmetric broadcast channels include the derivation of diamond-norm bounds on the asymptotic convergence of quantum cloning to state estimation and the derivation of bounds on the amount of quantum information that can be jointly decoded by a group of k receivers at the output of a symmetric broadcast channel.

The connection between quantum estimation and cloning is an inspiring leitmotiv of Quantum Information Theory [1,2,3,4,5,6,7,8]. The main related question is: how well can we simulate cloning via estimation? Or, more precisely, how well can we simulate cloning with a "measure-and-prepare" protocol where the input systems are measured, and the output systems are prepared in some state depending on the measurement outcome? As a particular instance of this question, one can ask whether "asymptotic cloning is state estimation" [9], that is, whether the gap between the single-particle fidelity of an optimal cloning channel and the fidelity of the corresponding optimal estimation vanishes when the number of clones tends to infinity.

In Ref. [7] Bae and Acín showed that a channel producing an infinite number of indistinguishable clones must be of the measure-and-prepare form. On the other hand, Ref. [8] showed that a channel producing a finite number $M < \infty$ of indistinguishable clones can be simulated by a measure-and-prepare channel introducing an error at most of order $\mathcal{O}(1/M)$ on each clone. The proof of

W. van Dam et al. (Eds.): TQC 2010, LNCS 6519, pp. 9–25, 2011.

Ref. [8] was based on the so-called *finite quantum de Finetti theorem* [10,11,12], that states that the restriction to k particles of a permutationally invariant M-partite state can be approximated with an error at most of order $\mathcal{O}(k/M)$ by a mixture of product states of the form $\rho^{\otimes k}$. This theorem represents the finite version of the *quantum de Finetti theorem* proved by Caves, Fuchs, and Schack [13] in the context of the Bayesian interpretation of quantum theory. The quantum de Finetti theorem of Ref. [13] corresponds to the ideal $M = \infty$ case and can be directly seen as the quantum formulation of the celebrated de Finetti theorem [14].

Apparently, finite quantum de Finetti theorems are the key to prove the equivalence between asymptotic cloning and estimation. The first result of this paper is to show that, in a sense, the converse is also true: a finite quantum de Finetti-type result can be derived from a particular relation between the optimal estimation [15,3] and the optimal cloning [2] of an unknown pure state. Precisely, we will see that the optimal measure-and-prepare channel sending M copies of an unknown pure states to k approximate copies is equivalent to a random loss of all but s particles followed by universal cloning from s to k copies. For $M >> k$ the term with $s = k$ dominates, implying that the optimal measure-and-prepare channel is close to the partial trace over all but k particles. As we will see, this implies directly a de Finetti-type result. Qualitatively, this result shows that the working principle of the finite de Finetti theorems is simply the fact that state estimation from M input copies to k output copies becomes almost perfect when M is large compared to k. Quantitatively, however, the bound derived from the representation of the optimal measure-and-prepare channel as a random mixture of losses followed by cloning can be tightened, as mentioned in subsection 1.4. The bound can be used to derive a finite de Finetti theorem for symmetric quantum broadcast channels, i.e. for channels that distribute quantum information to M indistinguishable users. Examples of symmetric broadcast channels are the channels for the optimal cloning of an unknown state ρ_i randomly drawn with probability p_i from some set of states $\{\rho_i\}$ [16]. The paper concludes with two applications of the finite de Finetti theorem for symmetric broadcast channels. First, the theorem will be used to provide diamond-norm bounds on the asymptotic convergence of quantum cloning to state estimation, thus strengthening the proof of Ref. [8]. As a second application, the theorem will be used to show that the restriction to k users of any symmetric broadcast channel has a quantum capacity that vanishes at rate $\mathcal{O}(k/M)$ in the large M asymptotics. Even if the overall channel is unitary, and therefore its capacity has the maximum possible value, a group of $k << M$ users will only be able to decode a vanishingly small amount of quantum information.

1   The Universal Measure-and-Prepare Channel

Let us start with some simple facts about the optimal measure-and-prepare channel transforming M copies of a completely unknown pure states into k approximate copies. The optimal quantum measurement for the estimation of a completely unknown pure state $|\psi\rangle \in \mathcal{H} \simeq \mathbb{C}^d$ from M input copies is given by the *coherent-state POVM* [15,3]

$$P_\varphi^{(M)} \, \mathrm{d}\varphi = d_+^{(M)} \, |\varphi\rangle\langle\varphi|^{\otimes M} \, \mathrm{d}\varphi \qquad d_+^{(M)} = \binom{d+M-1}{M} \tag{1}$$

where $|\varphi\rangle \in \mathcal{H}$ is a unit vector and $\mathrm{d}\varphi$ is the normalised $SU(d)$-invariant measure on pure states. This measurement provides a resolution of the identity in the symmetric subspace $(\mathcal{H}^{\otimes M})_+ \subseteq \mathcal{H}^{\otimes M}$, namely in the subspace spanned by the unit vectors

$$|\boldsymbol{n}\rangle := \frac{1}{\sqrt{M! n_1! n_2! \ldots n_d!}} \sum_{\pi \in S_M} U_\pi^{(M)} |1\rangle^{\otimes n_1} |2\rangle^{\otimes n_2} \ldots |d\rangle^{\otimes n_d} \tag{2}$$

where $|1\rangle, |2\rangle, \ldots, |d\rangle$ is a fixed orthonormal basis for $\mathcal{H}$, $\boldsymbol{n} = (n_1, n_2, \ldots, n_d)$ is a partition of M, the sum runs over the symmetric group S_M of all permutations of M objects, and $U_\pi^{(M)}$ is the unitary operator that permutes the M copies of $\mathcal{H}$ according to the permutation $\pi \in S_M$.

Denoting by $\mathcal{P}_{M,d}$ the set of partitions of M in d nonnegative integers, the normalization of the coherent-state POVM in Eq. (1) is given by

$$\int \mathrm{d}\varphi \, P_\varphi^{(M)} = \sum_{\boldsymbol{n} \in \mathcal{P}_{M,d}} |\boldsymbol{n}\rangle\langle\boldsymbol{n}| = P_+^{(M)}, \tag{3}$$

where $P_+^{(M)}$ is the projector on the symmetric subspace $(\mathcal{H}^{\otimes M})_+$.

We now consider the *universal measure-and-prepare channel* from M to k copies, namely the channel that measures the coherent-state POVM $P_\varphi^{(M)}$ and, according to the estimate, prepares k copies of the state $|\varphi\rangle$:

$$\mathcal{U}MeasPrep_{M,k}(\rho) := \int \mathrm{d}\varphi \, \mathrm{Tr}[P_\varphi^{(M)} \rho] \, |\varphi\rangle\langle\varphi|^{\otimes k}. \tag{4}$$

Using Eq. (3) with the substitution $M \to M+k$ one obtains the equivalent expression

$$\mathcal{U}MeasPrep_{M,k}(\rho) = d_+^{(M)} \int \mathrm{d}\varphi \, \mathrm{Tr}_M \left[(\rho \otimes I^{\otimes k}) \, |\varphi\rangle\langle\varphi|^{\otimes M+k} \right]$$

$$= \frac{d_+^{(M)}}{d_+^{(M+k)}} \mathrm{Tr}_M \left[(\rho \otimes I^{\otimes k}) \, P_+^{(M+k)} \right] \tag{5}$$

where Tr_M denotes the partial trace over the first M Hilbert spaces.

For an arbitrary pure state $|\psi\rangle$, the fidelity between the channel output $\mathcal{U}MeasPrep_{M,k}(|\psi\rangle\langle\psi|^{\otimes M})$ and the desideratum $|\psi\rangle\langle\psi|^{\otimes k}$ is given by $F_{M,k} = d_+^{(M)}/d_+^{(M+k)}$, as it is immediate from Eq. (5). In fact, it is easy to show that $F_{M,k} = d_+^{(M)}/d_+^{(M+k)}$ is the maximum average fidelity achievable with a measure-and-prepare channel $\mathcal{M}(\rho) = \sum_i \mathrm{Tr}[P_i \rho] \rho_i$, where $\{P_i\}$ is a POVM on $(\mathcal{H}^{\otimes M})_+$ and $\{\rho_i\}$ is a set of states on $(\mathcal{H}^{\otimes k})_+$. Indeed, in this case one has

$$\overline{F} = \int d\psi \langle \psi|^{\otimes k} \mathcal{M} \left(|\psi\rangle\langle\psi|^{\otimes M}\right) |\psi\rangle^{\otimes k} = \frac{\sum_i \mathrm{Tr}\left[(P_i \otimes \rho_i)P_+^{(M+k)}\right]}{d_+^{(M+k)}}$$

$$\leq \frac{\sum_i \mathrm{Tr}\left[P_i \otimes \rho_i\right]}{d_+^{(M+k)}} = \frac{d_+^{(M)}}{d_+^{(M+k)}}$$

(cf. Bruß and Macchiavello [3] for the $k = 1$ case). Clearly, when M is large compared to k the fidelity $F_{M,k}$ is close to unit: the desired output states $|\psi\rangle^{\otimes k}$ are much less distinguishable than the input states $|\psi\rangle^{\otimes M}$, thus allowing for an almost ideal re-preparation. In this case, one has

$$\mathcal{U}MeasPrep_{M,k}\left(|\psi\rangle\langle\psi|^{\otimes M}\right) \approx |\psi\rangle\langle\psi|^{\otimes k} \qquad \forall|\psi\rangle \in \mathcal{H},$$

or, equivalently (cf. the Appendix),

$$\mathcal{U}MeasPrep_{M,k}(\rho) \approx \mathrm{Tr}_{M-k}[\rho] \qquad \forall\rho \in \mathsf{Lin}\left(\left(\mathcal{H}^{\otimes M}\right)_+\right),$$

where $\mathsf{Lin}(V)$ denotes the set of linear operators on the linear space V ($V = \left(\mathcal{H}^{\otimes M}\right)_+$ in this case). Despite the simplicity of the above observation, the consequences of the fact that for $M >> k$ the estimation from M to k copies is "almost ideal" are far from trivial: as we will see, this simple fact can be considered as the working principle of the finite de Finetti theorems.

The purpose of the next subsection is to give a convenient representation of the channel $\mathcal{U}MeasPrep_{M,k}$ as a convex mixture of losses concatenated with cloning channels. Using this representation we will show that in the limit $k/M \rightarrow 0$ the channel $\mathcal{U}MeasPrep_{M,k}$ converges to the partial trace Tr_{M-k} in the strongest possible sense, in terms of the *diamond norm* [17], equivalent to the *norm of complete boundedness* [18] of the channel in Heisenberg picture. Operationally, convergence in the diamond norm means that for $M >> k$ the two channels $\mathcal{U}MeasPrep_{M,k}$ and Tr_{M-k} are almost indistinguishable even when entanglement-assisted discrimination strategies are employed.

1.1   Representation of the Universal Measure-and-Prepare Channel as a Mixture of Universal Cloning Channels

The main result of this subsection is the following expression, proved in the Appendix:

$$\mathcal{U}MeasPrep_{M,k}(\rho) = \sum_{s=0}^{\min\{k,M\}} p_s \,\mathcal{U}Clon_{s,k}\left(\mathrm{Tr}_{M-s}[\rho]\right), \quad p_s = \frac{\binom{M}{s}\binom{d+k-1}{k-s}}{\binom{d+M+k-1}{k}},$$

$$(6)$$

$\mathcal{U}Clon_{s,k}$ being the *universal s-to-k cloning channel*, i.e. the optimal quantum channel that clones an unknokwn pure state $|\psi\rangle$ from s to k copies, given by [2,4]

$$\mathcal{U}Clon_{s,k}(\rho) = \frac{d_+^{(s)}}{d_+^{(k)}} P_+^{(k)} \left(\rho \otimes I^{\otimes (k-s)} \right) P_+^{(k)}. \tag{7}$$

Note that $\{p_s\}$ is a probability distribution, as the normalization

$$\sum_{s=0}^{\min\{k,M\}} p_s = \sum_{s=0}^{k} p_s = 1$$

follows immediately from the fact that $p_s = 0$ if $s > M$ and from the Chu-Vandermonde convolution formula (see Eq. (7.6) p. 59 of Ref. [19] for an equivalent formula)

$$\binom{z+w}{N} = \sum_{i=0}^{N} \binom{z}{i} \binom{w}{N-i} \qquad \forall z, w \in \mathbb{C}, \forall N \in \mathbb{N}. \tag{8}$$

Eq. (6) means that measuring M copies and re-preparing k copies has the same effect of a random loss of $M - s$ systems followed by quantum cloning from s to k copies: the particles that are missing are replaced by clones.

In the following we will consider the two extreme cases $k >> M$ and $M >> k$. In the former, we will see that the measure-and-prepare channel $\mathcal{U}MeasPrep_{M,k}$ converges to the universal cloning $\mathcal{U}Clon_{M,k}$. In the latter, the measure-and-prepare channel $\mathcal{U}MeasPrep_{M,k}$ will converge to the partial trace Tr_{M-k}, leading to a de Finetti-type result. The convergence will be quantified in terms of the *diamond norm* [17] (in Heisenberg picture, the completely bounded norm [18]), which for a Hermitian-preserving map Δ from $\mathsf{Lin}(\mathcal{H}_{in})$ to $\mathsf{Lin}(\mathcal{H}_{out})$ is given by

$$\|\Delta\|_\diamond = \sup_{\mathcal{H}_A} \sup_{|\Psi\rangle \in \mathcal{H}_A \otimes \mathcal{H}_{in}, \|\Psi\|=1} \|(\mathcal{I}_A \otimes \Delta)(|\Psi\rangle\langle\Psi|)\|_1, \tag{9}$$

where $\|A\|_1 = \mathrm{Tr}|A|$ is the trace-norm and $\mathcal{I}_A$ is the identity map on the ancillary Hilbert space $\mathcal{H}_A$.

1.2 $k >> M$ Case: Convergence to Universal Cloning

Suppose that the number of output copies k is larger than the number of input copies M. In the limit of $M/k \to 0$, the term with $s = M$ in Eq. (6) dominates, thus giving $\mathcal{U}MeasPrep_{M,k} \approx \mathcal{U}Clon_{M,k}$.

An estimate of the diamond-norm convergence to universal cloning is given by the following:

Theorem 1 (Convergence to universal cloning). *The universal measure-and-prepare channel* $\mathcal{U}MeasPrep_{M,k}$ *converges to the universal cloning channel* $\mathcal{U}Clon_{M,k}$ *in the limit* $k \to \infty$. *In particular, the following bound holds:*

$$\left\| \mathcal{U}MeasPrep_{M,k} - \mathcal{U}Clon_{M,k} \right\|_\diamond \le \frac{2M(d+M-1)}{k+d}. \tag{10}$$

Proof. Writing $\mathcal{U}MeasPrep_{M,k} = p_M \mathcal{U}Clon_{M,k} + (1-p_M)\mathcal{R}est$ where $\mathcal{R}est$ is a suitable channel, one has $\|\mathcal{U}MeasPrep_{M,k} - \mathcal{U}Clon_{M,k}\|_\diamond \le (1-p_M)\|\mathcal{R}est - \mathcal{U}Clon_{M,k}\|_\diamond$. Since the distance between the two channels $\mathcal{R}est$ and $\mathcal{U}Clon_{M,k}$ is upper bounded by 2, this gives $\|\mathcal{U}MeasPrep_{M,k} - \mathcal{U}Clon_{M,k}\|_\diamond \le 2(1-p_M)$. The bound in Eq. (11) just comes from a lower bound on p_M:

$$
\begin{aligned}
p_M &= \frac{k(k-1)\dots(k-M+1)}{(d+M+k-1)(d+M+k-2)\dots(d+k)} \ge \left(\frac{k-M+1}{d+k}\right)^M \\
&= \left(1 - \frac{d+M-1}{d+k}\right)^M \ge 1 - \frac{M(d+M-1)}{d+k}.
\end{aligned}
$$

$\square$

Theorem 1 shows an exceptionally strong case of equivalence between asymptotic cloning and state estimation: it shows that, in the universal case, the optimal cloning channel [2,4] converges in diamond norm to the measure-and-prepare channel $\mathcal{U}MeasPrep_{M,k}$ when the number k of output copies is large with respect to the number M of input copies. It is worth stressing, however, that this result is very specific to the universal case. What can be proved for generic (i.e. non-universal) cloning channels is that the k-particle restrictions of a cloning channel with M output copies can be simulated by a measure-and-prepare channel with an error of order k/M (see subsection 2.2). This result will emerge from the analysis of Eq. (6) in the $M \gg k$ case, which is discussed in the next subsection.

1.3 $M \gg k$ Case: Convergence to the Partial Trace

Here we consider the case where the number is input copies k is large with respect to the number of output copies M. In this case, the leading term in Eq. (6) is the term with $s = k$. Note that, since for $s = k$ the universal cloning $\mathcal{U}Clon_{k,k}$ is simply the identity map on $(\mathcal{H}^{\otimes k})_+$, the corresponding term in Eq. (6) is the partial trace Tr_{M-k}. Therefore, when M is large compared to k the channel $\mathcal{U}MeasPrep_{M,k}$ converges to the trace Tr_{M-k}. This implies an almost ideal estimation, with $\mathcal{U}MeasPrep_{M,k}(|\psi\rangle\langle\psi|^{\otimes M}) \approx \mathrm{Tr}_{M-k}[|\psi\rangle\langle\psi|^{\otimes M}] = |\psi\rangle\langle\psi|^{\otimes k}$. A first estimate on the diamond-norm convergence to ideal estimation is given by the following

Theorem 2 (Convergence to ideal estimation). *The universal measure-and-prepare channel $\mathcal{U}MeasPrep_{M,k}$ converges to the trace channel Tr_{M-k} in the limit $M \to \infty$. In particular, the following bound holds*

$$
\|\mathcal{U}MeasPrep_{M,k} - \mathrm{Tr}_{M-k}\|_\diamond \le \frac{2k(d+k-1)}{M+d}. \tag{11}
$$

Proof. Writing $\mathcal{U}MeasPrep_{M,k} = p_k \mathrm{Tr}_{M-k} + (1-p_k)\mathcal{R}est$ where $\mathcal{R}est$ is a suitable channel, one has $\|\mathcal{U}MeasPrep_{M,k} - \mathrm{Tr}_{M-k}\|_\diamond \le (1-p_k)\|\mathcal{R}est - \mathrm{Tr}_{M-k}\|_\diamond$. Since the distance between the two channels $\mathcal{R}est$ and Tr_{M-k} is upper bounded by 2, this gives $\|\mathcal{U}MeasPrep_{M,k} - \mathrm{Tr}_{M-k}\|_\diamond \le 2(1-p_k)$. The bound in Eq. (11) just comes from a lower bound on p_k:

$$p_k = \frac{M(M-1)\ldots(M-k+1)}{(d+M+k-1)(d+M+k-2)\ldots(d+M)} \geq \left(\frac{M-k+1}{d+M}\right)^k$$

$$= \left(1 - \frac{d+k-1}{d+M}\right)^k \geq 1 - \frac{k(d+k-1)}{d+M}.$$

□

The bound of Eq. (11) clearly implies a de Finetti-type result:

Corollary 1. *For every state ρ with support in the symmetric space $(\mathcal{H}^{\otimes M})_+$ there exists a state $\tilde{\rho} = \sum_i p_i |\psi_i\rangle\langle\psi_i|^{\otimes M}$ such that the k-particle restrictions of ρ and $\tilde{\rho}$ are almost indistinguishable for large M. Precisely, denoting the k-particle restrictions by $\rho^{(k)} = \mathrm{Tr}_{M-k}[\rho]$ and $\tilde{\rho}^{(k)} = \mathrm{Tr}_{M-k}[\tilde{\rho}]$, one has*

$$\left\|\rho^{(k)} - \tilde{\rho}^{(k)}\right\|_1 \leq \frac{2k(d+k-1)}{M+d} \tag{12}$$

Proof. Taking $\tilde{\rho} = \mathcal{U}MeasPrep_{M,M}(\rho)$ we obtain a state of the desired form, and, in addition, we have

$$\left\|\tilde{\rho}^{(k)} - \rho^{(k)}\right\|_1 = \left\|\mathcal{U}MeasPrep_{M,k}(\rho) - \mathrm{Tr}_{M-k}[\rho]\right\|_1$$
$$\leq \left\|\mathcal{U}MeasPrep_{M,k} - \mathrm{Tr}_{M-k}\right\|_\diamond$$
$$\leq \frac{2k(d+k-1)}{M+d}.$$

□

The bound of Eq. (12) can be extended to the case of states on $\mathcal{H}^{\otimes M}$ that are just permutationally invariant, using the fact that *i)* every permutationally invariant state on $\mathcal{H}^{\otimes M}$ has a purification in the symmetric space $(\mathcal{K}^{\otimes M})_+$, with $\mathcal{K} = \mathcal{H} \otimes \mathcal{H}$ (see e.g. [10]) and that *ii)* the norm is non-increasing under partial traces. Therefore, for a permutationally invariant state the bound of Eq. (12) holds with the substitution $d \to d^2$.

1.4   Improving the Bound

The bound of Eq. (11) provides good estimates for $k = 1$ or when d is large, so that $Mk \leq d^2$ (see the observation below). Outside this range of values, the estimate can be improved using the technique developed in Ref. [10] for the proof of the finite de Finetti theorem, combined with the bounding of Ref. [8]:

Theorem 3. *The universal measure-and-prepare channel $\mathcal{U}MeasPrep_{M,k}$ satisfies the bound*

$$\left\|\mathcal{U}MeasPrep_{M,k} - \mathrm{Tr}_{M-k}\right\|_\diamond \leq 4\left(1 - \sqrt{\frac{d_+^{(M-k)}}{d_+^{(M)}}}\right) \leq \frac{2kd}{M} \tag{13}$$

Observation. Note that the quantity $2kd/M$ in Eq. (13) is larger than the quantity $2k(d+k-1)/(M+d)$ in Eq. (11) whenever $M(k-1) \leq d^2$. In general, the more accurate estimate is obtained by taking the minimum between the two quantities in Eqs. (11) and (13).

Proof of Theorem 2. Let $|\Psi\rangle$ be an arbitrary state in $\mathcal{H}_A \otimes \left(\mathcal{H}^{\otimes M}\right)_+$, where $\mathcal{H}_A$ is an arbitrary Hilbert space. Define the states

$$\rho^{(Ak)} = (\mathcal{I}_A \otimes \mathrm{Tr}_{M-k}) [|\Psi\rangle\langle\Psi|]$$
$$\tilde{\rho}^{(Ak)} = (\mathcal{I}_A \otimes \mathcal{U}MeasPrep_{M,k}) [|\Psi\rangle\langle\Psi|].$$

Using the normalization of the coherent-state POVM in Eq. (3) with the substitution $M \to M - k$, we can write $\rho^{(Ak)} = \int \mathrm{d}\varphi \, \rho_\varphi^{(Ak)}$, where

$$\rho_\varphi^{(Ak)} = \mathrm{Tr}_{M-k} \left[|\Psi\rangle\langle\Psi| \left(I_A \otimes I^{\otimes k} \otimes P_\varphi^{(M-k)}\right)\right] \, .$$

On the other hand, the state $\tilde{\rho}^{(Ak)}$ can be written as

$$\tilde{\rho}^{(Ak)} = \lambda \int \mathrm{d}\varphi \, \left(I_A \otimes P_\varphi^{(k)}\right) \rho_\varphi^{(Ak)} \left(I_A \otimes P_\varphi^{(k)}\right),$$

with $\lambda = \frac{d_+^{(M)}}{d_+^{(M-k)}d_+^{(k)2}}$. The difference between $\rho^{(Ak)} - \tilde{\rho}^{(Ak)}$ is then given by

$$\rho^{(Ak)} - \tilde{\rho}^{(Ak)} = \int \mathrm{d}\varphi \, (A_\varphi - B_\varphi A_\varphi B_\varphi),$$

where $A_\varphi = \rho_\varphi^{(Ak)}$ and $B_\varphi = \sqrt{\lambda} \left(I_A \otimes P_\varphi^{(k)}\right)$.

Using the relation $A - BAB = A(I - B) + (I - B)A - (I - B)A(I - B)$ we obtain

$$\rho^{(Ak)} - \tilde{\rho}^{(Ak)} = C + C^\dagger - D \, , \tag{14}$$

where $C = \int \mathrm{d}\varphi \, A_\varphi (I - B_\varphi)$ and $D = \int \mathrm{d}\varphi \, (I - B_\varphi) A_\varphi (I - B_\varphi)$. The operator C can be calculated using the relation

$$\int \mathrm{d}\varphi \, A_\varphi B_\varphi = \frac{\sqrt{\lambda}d_+^{(k)}d_+^{(M-k)}}{d_+^{(M)}} \int \mathrm{d}\varphi \, \mathrm{Tr}_{M-k} \left[|\Psi\rangle\langle\Psi| \left(I_A \otimes P_\varphi^{(M)}\right)\right]$$
$$= \sqrt{\frac{d_+^{(M-k)}}{d_+^{(M)}}} \, \mathrm{Tr}_{M-k}[|\Psi\rangle\langle\Psi|] = \sqrt{\frac{d_+^{(M-k)}}{d_+^{(M)}}} \rho^{(Ak)},$$

which gives $C = \left(1 - \sqrt{d_+^{(M-k)}/d_+^{(M)}}\right) \rho^{(Ak)} = C^\dagger$.

Taking the norm on both sides of Eq. (14), using the triangle inequality, and the fact that C and D are both nonnegative we obtain $\|\rho^{(Ak)} - \tilde{\rho}^{(Ak)}\|_1 \leq 2\|C\|_1 + \|D\|_1 = 2\mathrm{Tr}[C] + \mathrm{Tr}[D]$. Finally, taking the trace on both sides of Eq. (14) we get $\mathrm{Tr}[D] = 2\mathrm{Tr}[C]$. The inequality $\|\rho^{(Ak)} - \tilde{\rho}^{(Ak)}\|_1 \leq 4\mathrm{Tr}[C]$ then gives the first bound in Eq. (13). The second bound follows from the inequalities $d_+^{(M-k)}/d_+^{(M)} \geq (1 - k/M)^d$ (see e.g. Ref.[10]) and $(1-x)^\alpha \geq 1 - \alpha x$, which holds for $\alpha \geq 1$ and $x \leq 1$. $\square$

2   Symmetric Broadcast Channels

A *quantum broadcast channel* is a channel with a single sender and many receivers [20]. We define a *symmetric* broadcast channel as a channel where the Hilbert spaces of all receivers are isomorphic and the output of the channel is invariant under permutations. Precisely, we say that a channel $\mathcal{E} : \text{Lin}(\mathcal{H}_{in}) \rightarrow \text{Lin}\left(\mathcal{H}^{\otimes M}\right)$ is a *symmetric broadcast channel* if

$$\mathcal{E} = \mathcal{U}_\pi^{(M)} \mathcal{E} \qquad \forall \pi \in S_M, \tag{15}$$

where $\mathcal{U}_\pi^{(M)}$ is the unitary channel defined by $\mathcal{U}_\pi^{(M)}(\rho) := U_\pi^{(M)} \rho U_\pi^{(M)\dagger}$, $\rho \in \text{Lin}(\mathcal{H}_{in})$. The requirement of Eq. (15) models the situation where the quantum information in the input is equally spread over all receivers: any possible permutation of the receivers leaves the channel invariant. An example of symmetric broadcast channel is the optimal cloning channel for an arbitrary set of pure states, whenever the figure of merit is the average of the single-copy fidelity over all the M output copies (see e.g. [4]). In the following we will prove a finite de Finetti theorem for symmetric broadcast channels. The theorem is then used to show a strong form of the equivalence between asymptotic cloning and state estimation and to provide bounds on the amount of quantum information that can be jointly decoded by k receivers at the output of a symmetric broadcast channel.

2.1   Finite de Finetti Theorems for Symmetric Quantum Broadcast Channels

For symmetric broadcast channels with output in the symmetric subspace the following approximation result holds:

Theorem 4 (Finite de Finetti theorem for symmetric broadcast channels with output in the symmetric subspace). *For a symmetric broadcast channel* $\mathcal{E} : \text{Lin}(\mathcal{H}_{in}) \rightarrow \text{Lin}\left((\mathcal{H}^{\otimes M})_+\right)$ *there is a measure-and-prepare channel* $\widetilde{\mathcal{E}}$ *of the form* $\widetilde{\mathcal{E}}(\rho) = \sum_i \text{Tr}[P_i \rho] \, |\psi_i\rangle\langle\psi_i|^{\otimes M}$ *such that*

$$\|\widetilde{\mathcal{E}}^{(k)} - \mathcal{E}^{(k)}\|_\diamond \leq 4 \left(1 - \sqrt{\frac{d_+^{(M-k)}}{d_+^{(M)}}} \right) \leq \frac{2kd}{M}, \tag{16}$$

where $\widetilde{\mathcal{E}}^{(k)} := \text{Tr}_{M-k} \circ \widetilde{\mathcal{E}}$ *and* $\mathcal{E}^{(k)} := \text{Tr}_{M-k} \circ \mathcal{E}$.

Proof. Define the measure-and-prepare channel $\widetilde{\mathcal{E}}$ as

$$\widetilde{\mathcal{E}}(\rho) = \mathcal{U}MeasPrep_{M,M} \circ \mathcal{E}(\rho) = \int d\varphi \, \text{Tr}[Q_\varphi \rho] \, |\varphi\rangle\langle\varphi|^{\otimes M},$$

where $Q_\varphi d\varphi$ is the POVM defined by

$$\text{Tr}[Q_\varphi \rho] = \text{Tr}[P_\varphi^{(M)} \mathcal{E}(\rho)] \, \forall \rho \in \text{Lin}(\mathcal{H}_{in}),$$

that is, $Q_\varphi d\varphi$ is the POVM obtained by applying the channel $\mathcal{E}$ in Heisenberg picture to the coherent-state POVM $P_\varphi^{(M)} d\varphi$. From the definition of $\widetilde{\mathcal{E}}$ it is clear that $\mathcal{E}^{(k)} = \mathcal{U} MeasPrep_{M,k} \circ \mathcal{E}$. Using the submultiplicativity property $\|AB\|_\diamond \leq \|A\|_\diamond \|B\|_\diamond$, the fact that $\|\mathcal{E}\|_\diamond = 1$ since $\mathcal{E}$ is a channel, and the bound of Eq. (13) we then obtain

$$\left\|\widetilde{\mathcal{E}}^{(k)} - \mathcal{E}^{(k)}\right\|_\diamond = \|(\mathcal{U} MeasPrep_{M,k} - \text{Tr}_{M-k}) \circ \mathcal{E}\|_\diamond$$

$$\leq 4\left(1 - \sqrt{\frac{d_+^{(M)}}{d_+^{(M+k)}}}\right) \leq \frac{2dk}{M}. \qquad \square$$

The extension to arbitrary broadcast channels with permutationally invariant output is given in the following

Theorem 5 (Finite de Finetti theorem for symmetric broadcast channels). *For every symmetric broadcast channel $\mathcal{E} : \text{Lin}(\mathcal{H}_{in}) \to \text{Lin}\left(\mathcal{H}^{\otimes M}\right)$ there is a measure-and-prepare channel $\widetilde{\mathcal{E}} = \sum_i \text{Tr}[P_i \rho] \rho_i^{\otimes M}$ such that the bounds in Eq. (16) hold with the substitution $d \to d^2$.*

Proof. Consider the Stinespring dilation $\mathcal{E}(\rho) = \text{Tr}_{env}[V \rho V^\dagger]$, where $V : \mathcal{H}_{in} \to \mathcal{H}^{\otimes M} \otimes \mathcal{H}_{env}$ is an isometry and Tr_{env} is the partial trace over the environment Hilbert space $\mathcal{H}_{env}$. Since by definition a symmetric broadcast channel satisfies the relation

$$\mathcal{E}(\rho) = U_\pi^{(M)} \mathcal{E}(\rho) U_\pi^{(M)}, \qquad \forall \rho \in \text{Lin}(\mathcal{H}_{in}), \forall \pi \in S_M,$$

it follows from the theory of covariant channels that one can choose $\mathcal{H}_{env} = \mathcal{H}^{\otimes M} \otimes \mathcal{H}_{in}$ and V with the property

$$\left(U_\pi^{(M)} \otimes U_\pi^{(M)} \otimes I_{in}\right) V = V, \qquad \forall \pi \in S_M$$

(see Eq. (65) of Ref. [21]). This property implies that the output of the isometric channel $\mathcal{V}(\rho) = V \rho V^\dagger$ has support in the subspace $\left(\mathcal{K}^{\otimes M}\right)_+ \otimes \mathcal{H}_{in}$, where $\mathcal{K} = \mathcal{H}^{\otimes 2}$. Now, consider the channel $\mathcal{F} = \text{Tr}_{in} \circ \mathcal{V} : \text{Lin}(\mathcal{H}_{in}) \to \text{Lin}\left((\mathcal{K}^{\otimes M})_+\right)$. By theorem 4, there exists a measure-and-prepare channel $\widetilde{\mathcal{F}}$ of the form $\widetilde{\mathcal{F}}(\rho) = \sum_i \text{Tr}[P_i \rho] |\Psi_i\rangle\langle\Psi_i|^{\otimes M}$, with $|\Psi_i\rangle \in \mathcal{H}^{\otimes 2}$, such that the restrictions $\mathcal{F}^{(k)}$ and $\widetilde{\mathcal{F}}^{(k)}$ satisfy the bound of Eq. (16) with the substitution $d \to d^2$. To obtain the desired result it is sufficient to define the channel $\widetilde{\mathcal{E}}$ as $\widetilde{\mathcal{E}}(\rho) = \text{Tr}_{env}[\widetilde{\mathcal{V}}(\rho)] = \sum_i \text{Tr}[P_i \rho] \rho_i^{\otimes M}$, where ρ_i is the reduced density matrix of $|\Psi_i\rangle\langle\Psi_i|$, and to use the relation

$$\|\widetilde{\mathcal{E}}^{(k)} - \mathcal{E}^{(k)}\|_\diamond = \|\text{Tr}_{env,k} \circ (\widetilde{\mathcal{F}}^{(k)} - \mathcal{F}^{(k)})\|_\diamond$$

$$\leq \|\widetilde{\mathcal{F}}^{(k)} - \mathcal{F}^{(k)}\|_\diamond,$$

where $\text{Tr}_{env,k}$ denotes the partial trace over the k systems in the environment.

$\square$

Observation. The usual de Finetti theorems for quantum states [10,11,12] can be retrieved from theorems 4 and 5 in the special case of symmetric broadcasting channels with trivial input space $\mathcal{H}_{in} \simeq \mathbb{C}$. In this case the POVM $\{P_i\}$ becomes just a collection of probabilities $\{p_i\}$.

Theorems 4 and 5 have many interesting consequences: first of all they imply that the output state of k receivers contains a vanishing amount of entanglement in the limit of vanishing k/M. Moreover, they imply that the information transmitted to a small number of receivers can only be classical, while the amount of quantum information is vanishing. This observation will be made quantitatively precise in subsection 2.3. Another consequence is a strong form of the equivalence between asymptotic cloning states estimation, briefly discussed in the next subsection.

2.2 Strong Equivalence between Asymptotic Pure State Cloning and State Estimation

Let $\{|\psi_x\rangle\}_{x\in X} \subset \mathcal{H}$ be a set of pure states and $\{p_x\}$ a corresponding set of prior probabilities. An N-to-M cloning channel transforms N copies of a state $|\psi_x\rangle$ into M approximate copies, the joint state of the copies being a state on $\mathcal{H}^{\otimes M}$. The requirement that each single copy have the same fidelity with the state $|\psi_x\rangle$ is implemented without loss of generality by taking cloning channels with permutationally invariant output: clearly, such cloning channels are an example of symmetric broadcast channels. Let us call $Clon_{N,M}$ the N-to-M cloning channel under consideration and let $\widetilde{Clon}_{N,M}$ be the measure-and-prepare channel defined in Theorem 5. Theorem 5 then implies the bound

$$\left\| Clon_{N,M}^{(k)} - \widetilde{Clon}_{N,M}^{(k)} \right\|_{\diamond} \leq \frac{2d^2 k}{M}, \tag{17}$$

that is, for fixed k and d the cloning channel becomes more and more indistinguishable from a measure-and-prepare channel as M increases. In particular, if $Clon_{N,M}$ is the optimal cloning channel according to some figure of merit, Eq. (17) entails the convergence of optimal cloning to estimation. Note that the convergence in diamond norm represents an improvement over the trace-norm convergence of Ref. [8], as it states that cloning is indistinguishable from estimation even with the aid of entanglement with a reference system. The convergence of the fidelities is then a simple corollary: For every state ψ_x, the single-copy fidelity is given by

$$F_{clon}[N, M, x] = \langle \psi_x | Clon_{N,M}^{(1)}(|\psi_x\rangle\langle\psi_x|^{\otimes M}) | \psi_x \rangle.$$

Denoting by $F_{\widetilde{clon}}[N, x]$ the single-copy fidelity for the measure-and-prepare channel $\widetilde{Clon}_{N,M}$ (note that in this case the fidelity is independent of M), we have

$$|F_{clon}[N, M, x] - F_{\widetilde{clon}}[N, x]| \leq \left\| (Clon_{N,M}^{(1)} - \widetilde{Clon}_{N,M}^{(1)})(|\psi_x\rangle\langle\psi_x|^{\otimes N}) \right\|_1$$

$$\leq \left\| Clon_{N,M}^{(1)} - \widetilde{Clon}_{N,M}^{(1)} \right\|_{\diamond} \leq \frac{2d^2 k}{M}.$$

Denoting by $F_{est}[N]$ the maximum average fidelity achievable by a measure-and-prepare channel and using the fact that $F_{est}[N] \leq F_{clon}[N, M], \forall M$ we then have the bound

$$0 \leq F_{clon}[N, M] - F_{est}[N] \leq \left| \sum_x p_x (F_{clon}[N, M, x] - F_{\widetilde{clon}}[N, x]) \right|$$

$$\leq \sum_x p_x \left| F_{clon}[N, M, x] - F_{\widetilde{clon}}[N, x] \right| \leq \frac{2d^2 k}{M},$$

which implies the limit $\lim_{M \to \infty} F_{clon}[N, M] = F_{est}[N]$.

2.3 Bounds on the Quantum Capacities of the k-Receivers Restriction of a Symmetric Broadcast Channel

Theorems 4 and 5 also imply a set of bounds on the amount of quantum information that k receivers can jointly decode at the output of a symmetric broadcast channel $\mathcal{E}$. For definiteness, let us consider the case of a channel $\mathcal{E}$ with output in the symmetric subspace $(\mathcal{H}^{\otimes M})_+$: this is the case, e.g. of all known examples of optimal pure state cloning [16]. A first bound on the quantum capacity comes from the continuity result of Ref.[22], that, along with the fact that measure-and-prepare channels have zero quantum capacity, yields the following estimate

$$Q(\mathcal{E}^{(k)}) = |Q(\mathcal{E}^{(k)}) - Q(\widetilde{\mathcal{E}}^{(k)})| \leq \frac{16kd}{M} \log d_+^{(k)} + 4H\left(\frac{2kd}{M}\right). \quad (18)$$

where H is the binary entropy $H(x) = -x \log x - (1-x) \log(1-x)$, and log denotes the logarithm in base 2.

Two other estimates are given in the following

Corollary 2. *The quantum capacity of the k-receivers restriction of a symmetric broadcast channel $\mathcal{E} : \mathrm{Lin}(\mathcal{H}_{in}) \to \mathrm{Lin}\left((\mathcal{H}^{\otimes M})_+\right)$ satisfies the bound*

$$Q(\mathcal{E}^{(k)}) \leq \min \left\{ \log\left(1 + \frac{2kdd_+^{(k)}}{M}\right), \log\left(1 + \frac{2kdd_{in}}{M}\right) \right\} \quad (19)$$

$$\leq \min \left\{ \frac{2kdd_+^{(k)}}{M}, \frac{2kdd_{in}}{M} \right\} \quad (20)$$

Proof. Holevo and Werner proved that the quantum capacity of a channel $\mathcal{C}$ is upper bounded by the ε-quantum capacity $Q_\varepsilon(\mathcal{C})$ [23] (i.e. the supremum of the rates that are asymptotically achievable with error bounded by ε), and that $Q_\varepsilon(\mathcal{C})$ is upper bounded by $\log \|\mathcal{C}\Theta_{in}\|_\diamond$, where Θ_{in} is the transposition map on the input space $\mathcal{H}_{in}$. We then obtain

$$Q(\mathcal{E}^{(k)}) \leq Q_\epsilon(\mathcal{E}^{(k)}) \leq \log \|\widetilde{\mathcal{E}}^{(k)}\Theta_{in} + (\mathcal{E}^{(k)} - \widetilde{\mathcal{E}}^{(k)})\Theta_{in}\|_\diamond$$

$$\leq \log\left(\|\widetilde{\mathcal{E}}^{(k)}\Theta_{in}\|_\diamond + \|\mathcal{E}^{(k)} - \widetilde{\mathcal{E}}^{(k)}\|_\diamond \|\Theta_{in}\|_\diamond\right)$$

$$\leq \log\left(1 + \frac{2kdd_{in}}{M}\right),$$

having used the triangle inequality, the submultiplicativity $\|\mathcal{AB}\|_\diamond \leq \|\mathcal{A}\|_\diamond \|\mathcal{B}\|_\diamond$ the fact that $\|\widetilde{\mathcal{E}}^{(k)}\Theta_{in}\|_\diamond = 1$ since $\widetilde{\mathcal{E}}^{(k)}\Theta_{in}(\rho) = \int d\varphi \mathrm{Tr}[Q_\varphi^T \rho]|\varphi\rangle\langle\varphi|^{\otimes k}$ is still a quantum channel, the equality $\|\Theta_{in}\|_\diamond = d_{in}$, and the bound of Eq. (16). Similarly, denoting by $\Theta_+^{(M)}$ and $\Theta_+^{(k)}$ the transposition maps on $\left(\mathcal{H}^{\otimes M}\right)_+$ and $\left(\mathcal{H}^{\otimes k}\right)_+$, respectively, we obtain

$$
\begin{aligned}
Q(\mathcal{E}^{(k)}) \leq Q_\epsilon(\mathcal{E}^{(k)}) &\leq \log \|\widetilde{\mathcal{E}}^{(k)}\Theta + (\mathcal{E}^{(k)} - \widetilde{\mathcal{E}}^{(k)})\Theta_{in}\|_\diamond \\
&\leq \log\left[1 + \|(\mathcal{U}MeasPrep_{M,k} - \mathrm{Tr}_{M-k})\Theta_+^{(M)}(\Theta_+^{(M)}\mathcal{E}\Theta_{in})\|_\diamond\right] \\
&\leq \log\left[1 + \|(\mathcal{U}MeasPrep_{M,k} - \mathrm{Tr}_{M-k})\Theta_+^{(M)}\|_\diamond\right] \\
&\leq \log\left[1 + \|\Theta_+^{(k)}\|_\diamond \|\Theta_+^{(k)}(\mathcal{U}MeasPrep_{M,k} - \mathrm{Tr}_{M-k})\Theta_+^{(M)}\|_\diamond\right] \\
&= \log\left[1 + \|\Theta_+^{(k)}\|_\diamond \|\mathcal{U}MeasPrep_{M,k} - \mathrm{Tr}_{M-k}\|_\diamond\right] \\
&\leq \log\left(1 + \frac{2kdd_+^{(k)}}{M}\right).
\end{aligned}
$$

having used the triangle inequality, the submultiplicativity $\|\mathcal{AB}\|_\diamond \leq \|\mathcal{A}\|_\diamond \|\mathcal{B}\|_\diamond$, the fact that $\Theta_+^{(M)}\mathcal{E}\Theta_{in}$ is a channel and that $\Theta_+^{(k)}\mathcal{U}MeasPrep_{M,k}\Theta_+^{(M)} = \mathcal{U}MeasPrep_{M,k}$ and $\Theta_+^{(k)}\mathrm{Tr}_{M-k}\Theta_+^{(M)} = \mathrm{Tr}_{M-k}$. The two bounds above prove Eq. (19). Eq. (20) then follows immediately from the relation $\log(1+x) \leq x$. □

Since the input quantum information has to be spread uniformly over a large number of receivers, a finite group of $k << M$ receivers can only access a vanishing amount of information. This fact holds even if the overall channel $\mathcal{E}$ is unitary (for example, if $\mathcal{E}$ is the identity channel from a super-user holding all input systems to M users, each of them receiving one output system).

3   Conclusions

In this paper we have seen that the standard finite quantum de Finetti theorems can be naturally rephrased as theorems about the diamond-norm distance between the optimal measure-and-prepare channel from M to k copies and the trace channel Tr_{M-k}. The working principle of the theorems appears to be the simple fact that estimation and re-preparation from M to k copies becomes almost ideal whenever M is large with respect to k. This idea suggests that similar approximation theorems could be obtained from other measure-and-prepare protocols based on estimation, where the input is given by M copies of some state $|\psi_x\rangle, x \in X$ and the goal is to produce k approximate copies. In this case, one can expect to obtain approximation theorems for multipartite quantum states in the linear span of the projectors $|\psi_x\rangle\langle\psi_x|^{\otimes M}$. The exploration of such generalizations is an interesting direction of future research.

Acknowledgements. I would like to thank D. Gottesman, R. Spekkens, I. Marvian, and A. Harrow for stimulating questions that helped me to improve the presentation. Research at Perimeter Institute is supported by the Government of Canada through Industry Canada and by the Province of Ontario through the Ministry of Research and Innovation.

References

1. Gisin, N., Massar, S.: Phys. Rev. Lett. 79, 2153 (1997)
2. Werner, R.F.: Phys. Rev. A 58, 1827 (1998)
3. Bruß, D., Ekert, A., Macchiavello, C.: Phys. Rev. Lett. 81, 2598 (1998)
4. Keyl, M., Werner, R.F.: J. Math. Phys. 40, 3283 (1999)
5. Bruß, D., Cinchetti, M., DAriano, G.M., Macchiavello, C.: Phys. Rev. A 62, 12302 (2000)
6. D'Ariano, G.M., Macchiavello, C.: Phys. Rev. A 67, 042306 (2003)
7. Bae, J., Acín, A.: Phys. Rev. Lett. 97, 30402 (2006)
8. Chiribella, G., D'Ariano, G.M.: Phys. Rev. Lett. 97, 250503 (2006)
9. Keyl, M.: Problem 22 of the list, http://www.imaph.tu-bs.de/qi/problems/
10. Christandl, M., Koenig, R., Mitchison, G., Renner, R.: Comm. Math. Phys. 273, 473 (2007)
11. Renner, R.: Nature Physics 3, 645 (2007)
12. Koenig, R., Mitchison, G.: J. Math. Phys. 50, 12105 (2009)
13. Caves, C.M., Fuchs, C.A., Schack, R.: J. Math. Phys. 43, 4537 (2002)
14. de Finetti, B.: Theory of Probability. Wiley, New York (1990)
15. Massar, S., Popescu, S.: Phys. Rev. Lett. 74, 1259 (1995)
16. Scarani, V., Iblisdir, S., Gisin, N., Acín, A.: Rev. Mod. Phys. 77, 1225 (2005)
17. Aharonov, D., Kitaev, A., Nisan, N.: Quantum Circuits with Mixed States. In: Proceedings of the 30th Annual ACM Symposium on Theory of Computing (STOC). ACM, New York (1998)
18. Paulsen, V.I.: Completely bounded maps and dilations. Longman Scientific and Technical (1986)
19. Askey, R.: Orthogonal polynomials and special functions, Philadelphia, PA. CBMS-NSF Regional Conference Series in Applied Mathematics, vol. 21 (1975)
20. Yard, J., Hayden, P., Devetak, I.: arXiv:quant-ph/0603098v1
21. Chiribella, G., D'Ariano, G.M., Perinotti, P.: J. Math. Phys. 50, 42101 (2009)
22. Leung, D., Smith, G.: Comm. Math. Phys. 292, 201 (2009)
23. Holevo, A.S., Werner, R.F.: Phys. Rev. A 3, 32312 (2001)
24. Klee, V.: Canad. J. Math. 16, 517 (1963)

Appendix

The Appendix is devoted to the derivation of Eq. (6). To this purpose we will use the fact that every operator $\rho \in \mathsf{Lin}\left((\mathcal{H}^{\otimes M})_+\right)$ can be written as a linear combination of the rank-one projectors $|\psi\rangle\langle\psi|^{\otimes M}$. An easy proof of this fact is given as follows: Let us write $|\psi\rangle = \sum_{k=1}^{d} \psi_k |k\rangle$. Then, we have (cf. Eq. 2 of Ref. [2])

$$|\psi\rangle^{\otimes M} = \sum_{\boldsymbol{n} \in \mathcal{P}_{M,d}} \psi_1^{n_1} \dots \psi_d^{n_d} \sqrt{\frac{M!}{n_1! \dots n_d!}} |\boldsymbol{n}\rangle,$$

and also

$$\frac{1}{M!} \left(\prod_{k=1}^{d} \frac{1}{\sqrt{m_k!}} \frac{\partial^{m_k}}{\partial \psi_k^{m_k}}\right) \left(\prod_{l=1}^{d} \frac{1}{\sqrt{n_l!}} \frac{\partial^{n_l}}{\partial \psi_k^{*n_l}}\right) |\psi\rangle\langle\psi|^{\otimes M} \Bigg|_{\psi=0} = |\boldsymbol{m}\rangle\langle\boldsymbol{n}|,$$

where the coefficients $\{\psi_k\}_{k=1}^{d}$ and their complex conjugates $\{\psi_l^*\}_{l=1}^{d}$ are treated as independent variables. This means that the operators $|\boldsymbol{m}\rangle\langle\boldsymbol{n}|$ are in the linear span of the projectors $|\psi\rangle\langle\psi|^{\otimes M}$ (indeed, the derivatives are limits of linear combinations, and, since we are in finite dimensions, any linear span is a closed set, containing all its limit points). Since the operators $\{|\boldsymbol{m}\rangle\langle\boldsymbol{n}|\}_{\boldsymbol{m},\boldsymbol{n}\in\mathcal{P}_{M,d}}$ span $\mathsf{Lin}\left((\mathcal{H}^{\otimes M})_+\right)$, the projectors $|\psi\rangle\langle\psi|^{\otimes M}$ also do. Note that the same conclusion would be obtained, through a lengthier calculation, by taking all possible derivatives with respect to the real parts $\{\mathsf{Re}(\psi_k)\}_{k=1}^{d}$ and the imaginary parts $\{\mathsf{Im}(\psi_k)\}_{k=1}^{d}$, instead of the derivatives with respect to the coefficients $\{\psi_k\}_{k=1}^{d}$ and their complex conjugates $\{\psi_k^*\}_{k=1}^{d}$.

Due to the above discussion, to prove Eq. (6) it is enough to characterize the action of $\mathcal{U}MeasPrep_{M,k}$ on a generic projector $|\psi\rangle\langle\psi|^{\otimes M}$. Moreover, since the choice of the basis $\{|1\rangle, |2\rangle, \dots, |d\rangle\}$ is arbitrary, for given $|\psi\rangle$ we can choose $|1\rangle = |\psi\rangle$. Then, Eq. (5) gives

$$\mathcal{U}MeasPrep_{M,k}(|1\rangle\langle1|^{\otimes M}) = \frac{d_+^{(M)}}{d_+^{(M+k)}} \sum_{\boldsymbol{m},\boldsymbol{n}\in\mathcal{P}_{k,d}} \alpha_{\boldsymbol{m},\boldsymbol{n}} |\boldsymbol{m}\rangle\langle\boldsymbol{n}|$$

with $\alpha_{\boldsymbol{m},\boldsymbol{n}} = \langle1|^{\otimes M}\langle\boldsymbol{m}| P_+^{(M+k)} |1\rangle^{\otimes M}|\boldsymbol{n}\rangle$. Using the relation

$$P_+^{(M+k)} = \frac{1}{(M+k)!} \sum_{\pi \in S_{M+k}} U_\pi^{(M+k)}$$

and Eq. (2) with the substitution $M \to k$, we obtain $\alpha_{\boldsymbol{m},\boldsymbol{n}} = \frac{k!(M+n_1)!}{(M+k)!n_1!}\delta_{\boldsymbol{m},\boldsymbol{n}}$, and, therefore,

$$\mathcal{U}MeasPrep_{M,k}(|1\rangle\langle1|^{\otimes M}) = \frac{d_+^{(M)}}{d_+^{(M+k)}} \binom{M+k}{k}^{-1} \sum_{\boldsymbol{n}\in\mathcal{P}_{k,d}} \binom{M+n_1}{M} |\boldsymbol{n}\rangle\langle\boldsymbol{n}|.$$

$$(21)$$

Using again Eq. (2) with the substitution $M \to k$ we get the chain of equalities

$$\sum_{\boldsymbol{n} \in \mathcal{P}_{k,d}} \binom{M+n_1}{M} |\boldsymbol{n}\rangle\langle\boldsymbol{n}| =$$

$$= \sum_{\boldsymbol{n} \in \mathcal{P}_{k,d}} \left(\frac{\binom{M+n_1}{M}}{k! n_1! \dots n_d!} \sum_{\pi, \sigma \in S_k} U_\pi^{(k)} \left(|1\rangle\langle 1|^{\otimes n_1} \otimes \cdots \otimes |d\rangle\langle d|^{\otimes n_d} \right) U_\sigma^{(k)} \right)$$

$$= \sum_{n_1=0}^{k} \frac{\binom{M+n_1}{M}}{k! n_1! (k-n_1)!} \sum_{\pi,\sigma \in S_k} U_\pi^{(k)} \left(|1\rangle\langle 1|^{\otimes n_1} \otimes (I - |1\rangle\langle 1|)^{\otimes(k-n_1)} \right) U_\sigma^{(k)}$$

$$= \sum_{n_1=0}^{k} \sum_{j=0}^{k-n_1} \frac{(-1)^j \binom{M+n_1}{M}\binom{k-n_1}{j}}{k! n_1! (k-n_1)!} \sum_{\pi,\sigma \in S_k} U_\pi^{(k)} \left(|1\rangle\langle 1|^{\otimes(n_1+j)} \otimes I^{\otimes(k-n_1-j)} \right) U_\sigma^{(k)}.$$

Defining $s = n_1 + j$, the chain can be continued as

$$\sum_{\boldsymbol{n} \in \mathcal{P}_{k,d}} \binom{M+n_1}{M} |\boldsymbol{n}\rangle\langle\boldsymbol{n}| =$$

$$= \sum_{n_1=0}^{k} \sum_{s=n_1}^{k} \frac{(-1)^{s-n_1} \binom{M+n_1}{M}\binom{k-n_1}{s-n_1}}{k! n_1! (k-n_1)!} \sum_{\pi,\sigma \in S_k} U_\pi^{(k)} \left(|1\rangle\langle 1|^{\otimes s} \otimes I^{\otimes(k-s)} \right) U_\sigma^{(k)}$$

$$= \sum_{s=0}^{k} \sum_{n_1=0}^{s} (-1)^{s-n_1} \binom{M+n_1}{M} \binom{k}{s} \binom{s}{n_1} P_+^{(k)} \left(|1\rangle\langle 1|^{\otimes s} \otimes I^{\otimes(k-s)} \right) P_+^{(k)}$$

Finally, we can use the combinatorial identity (see proof below)

$$\beta_s := \sum_{n=0}^{s} (-1)^{s-n} \binom{s}{n} \binom{M+n}{M} = \binom{M}{s} \tag{22}$$

to obtain

$$\sum_{\boldsymbol{n} \in \mathcal{P}_{k,d}} \binom{M+n_1}{M} |\boldsymbol{n}\rangle\langle\boldsymbol{n}| = \sum_{s=0}^{k} \binom{k}{s} \binom{M}{s} P_+^{(k)} (|1\rangle\langle 1|^{\otimes s} \otimes I^{\otimes k-s}) P_+^{(k)}. \tag{23}$$

Since $\binom{M}{s} = 0$ whenever $s > M$, the sum is in fact a sum from 0 to $\min\{M, k\}$. Combining Eqs. (21), (23), and (7) we obtain the expression

$$\mathcal{U}MeasPrep_{M,k}(|1\rangle\langle1|^{\otimes M}) = \sum_{s=0}^{\min\{k.M\}} \frac{d_+^{(M)} \binom{k}{s} \binom{M}{s}}{d_+^{(M+k)} \binom{M+k}{k}} P_+^{(k)}(|1\rangle\langle1|^{\otimes s} \otimes I^{\otimes k-s}) P_+^{(k)}$$

$$= \sum_{s=0}^{\min\{k,M\}} \frac{\binom{M}{s} \binom{d+k-1}{k-s}}{\binom{d+M+k-1}{k}} \mathcal{U}Clon_{s,k}(|1\rangle\langle1|^{\otimes s}),$$

which holds for arbitrary M and k, and for an arbitrary vector $|1\rangle = |\psi\rangle$. Hence, we have obtained Eq. (6).

Regarding the combinatorial identity of Eq. (22), it can be proved as follows: First, using Chu-Vandermonde formula (Eq. (8)) one obtains $\beta_s = \sum_{n=0}^{s} \sum_{l=0}^{M} (-1)^{s-n} \binom{s}{n} \binom{s+n}{l} \binom{M-s}{M-l}$ Then, Klee's identity (Proposition 1.1 of Ref. [24]) yields $\beta_s = \sum_{l=0}^{M} \binom{s}{l-s} \binom{M-s}{M-l} = \sum_{l'=0}^{M-s} \binom{s}{l'} \binom{M-s}{M-s-l'}$. Finally, the expression $\beta_s = \binom{M}{s}$ follows by applying Chu-Vandermonde formula again.

Simple Sets of Measurements for Universal Quantum Computation and Graph State Preparation

Yasuhiro Takahashi

NTT Communication Science Laboratories, NTT Corporation,
3-1 Morinosato-Wakamiya, Atsugi, Kanagawa 243-0198, Japan

Abstract. We show that the set of observables $\{Z \otimes X, (\cos\theta)X + (\sin\theta)Y$ all $\theta \in [0, 2\pi)\}$ with one ancillary qubit is universal for quantum computation. The set is simpler than a previous one in the sense that one-qubit projective measurements described by the observables in the set are ones only in the (X, Y) plane of the Bloch sphere. The proof of the universality implies a simple set of observables that is approximately universal for quantum computation. Moreover, it implies a simple set of observables for efficient graph state preparation.

1 Introduction

Measurement-based quantum computation uses only projective measurements for universal quantum computation in contrast to conventional models. There are many models for measurement-based quantum computation, such as cluster state computation [1] and teleportation-based quantum computation [2]. These models suggest a new way of realizing a quantum computer. Minimizing the resources required for universal quantum computation is important for realizing a quantum computer based on these models.

We consider the problem under the assumption that we can use only projective measurements and do not have initial cluster states [2,3,4]. The resources we focus on are observables, which describe projective measurements, and ancillary qubits. There have been many studies in this direction [3,4,5,6,7]. In 2005, Jorrand and Perdrix showed that the set of observables $\{Z \otimes X, Z, (\cos\theta)X + (\sin\theta)Y$ all $\theta \in [0, 2\pi)\}$ with one ancillary qubit is universal for quantum computation [6]. It has not been known whether a simpler universal set of observables can be constructed without increasing the number of ancillary qubits.

We show that the set of observables $\mathcal{S}_1 = \{Z \otimes X, (\cos\theta)X + (\sin\theta)Y$ all $\theta \in [0, 2\pi)\}$ with one ancillary qubit is universal. The set is simpler than Jorrand and Perdrix's [6] in the sense that one-qubit projective measurements described by the observables in $\mathcal{S}_1$ are ones only in the (X, Y) plane of the Bloch sphere. The key idea of the proof is to use Y-measurements appropriately in place of other one-qubit projective measurements. In contrast to the previous proof [6], our proof immediately implies the best known result for the approximate universality by Perdrix [7] that a set of two one-qubit observables and one two-qubit

W. van Dam et al. (Eds.): TQC 2010, LNCS 6519, pp. 26–34, 2011.

observable with one ancillary qubit is approximately universal. Such an example is the set of observables $\mathcal{S}_2 = \{Z \otimes X, Y, (X + Y)/\sqrt{2}\}$.

We also consider the problem of minimizing the resources required for preparing graph states efficiently. It is important to investigate this problem since graph states play a key role in quantum information processing. Høyer et al. showed that, for any graph $G = (V, E)$, some signed graph state $|G\rangle$ can be prepared by a quantum circuit consisting of one-qubit and two-qubit projective measurements with size $O(|V| + |E|)$, depth $O(|E|)$, and one ancillary qubit [8]. The circuit uses the set of observables $\{Z \otimes X, Z, X, (X - Y)/\sqrt{2}\}$.

Using the proof of the universality of $\mathcal{S}_1$, we show that the set of observables $\mathcal{S}_3 = \{Z \otimes X, Y\}$ with one ancillary qubit is sufficient for preparing graph states efficiently. More precisely, for any graph $G = (V, E)$, the (exact) graph state $|G\rangle$ can be prepared by a quantum circuit consisting of one-qubit and two-qubit projective measurements described by the observables in $\mathcal{S}_3$ with size and depth $O(|V|+|E|)$ and one ancillary qubit. The depth is $O(|E|)$ for the graphs in which we are interested. Though the usual method for preparing graph states performs controlled-Z operations, it is difficult to do so since $\mathcal{S}_3$ has only $Z \otimes X$ and Y. The key idea is to perform operations similar to controlled-Z operations and to remove the side effects of the similar operations by using Y-measurements.

2   Preliminaries

Pauli matrices X, Y, and Z are defined by

$$\begin{pmatrix} 0 & 1 \\ 1 & 0 \end{pmatrix}, \begin{pmatrix} 0 & -i \\ i & 0 \end{pmatrix}, \begin{pmatrix} 1 & 0 \\ 0 & -1 \end{pmatrix},$$

respectively. The observable Z describes the one-qubit projective measurement in the basis $\{|0\rangle, |1\rangle\}$ and the corresponding classical outcomes are 1 and -1, respectively. For any $\theta \in [0, 2\pi)$, the observable $(\cos\theta)X + (\sin\theta)Y$ describes the one-qubit projective measurement in the basis $\{|+_\theta\rangle, |-_\theta\rangle\}$, where

$$|+_\theta\rangle = \frac{|0\rangle + e^{i\theta}|1\rangle}{\sqrt{2}}, \quad |-_\theta\rangle = \frac{|0\rangle - e^{i\theta}|1\rangle}{\sqrt{2}}$$

and the corresponding classical outcomes are 1 and -1, respectively. This is a measurement in the (X, Y) plane of the Bloch sphere. We denote $|\pm_0\rangle$ as $|\pm\rangle$. Pauli matrices also denote unitary operations and we use σ_x, σ_y, and σ_z in the case. We also consider two-qubit observables such as $Z \otimes X$, where $\otimes$ denotes the tensor product. The projective measurement described by $Z \otimes X$ has only two possible classical outcomes 1 and -1. It consists of two projections: one is on the space spanned by $|0\rangle|+\rangle$ and $|1\rangle|-\rangle$ and the other is on the space spanned by $|0\rangle|-\rangle$ and $|1\rangle|+\rangle$.

Let $\mathcal{S}$ be a set of observables and U be a unitary operation. The simulation of U by using projective measurements described by the observables in $\mathcal{S}$ is decomposed into the following steps [7]:

1. Simulation step: σU is probabilistically implemented by using projective measurements described by the observables in $\mathcal{S}$, where σ is σ_x, σ_y, σ_z, or an identity operation I when U is on one qubit, and is known by the classical outcomes of the measurements. When U is on multiple qubits, σ is allowed to be a tensor product of these operations.

2. Correction step: If σU is implemented in the simulation step where $\sigma \neq I$, σ is implemented by using projective measurements described by the observables in $\mathcal{S}$ to obtain $\sigma\sigma U = U$.

In the standard quantum circuit model, a set of gates is universal for quantum computation if any unitary operation can be implemented exactly by a quantum circuit consisting only of gates in the set. The approximate universality of a set of gates is defined similarly. It is known that the set of all one-qubit gates and controlled-Z gate ΛZ are universal and that the set of Hadamard gate H, $\pi/8$ gate $Z(\pi/4)$, and ΛZ are approximately universal [9], where H, $Z(\theta)$, and ΛZ are defined by

$$\frac{1}{\sqrt{2}}\begin{pmatrix} 1 & 1 \\ 1 & -1 \end{pmatrix}, \begin{pmatrix} 1 & 0 \\ 0 & e^{i\theta} \end{pmatrix}, \begin{pmatrix} 1 & 0 & 0 & 0 \\ 0 & 1 & 0 & 0 \\ 0 & 0 & 1 & 0 \\ 0 & 0 & 0 & -1 \end{pmatrix},$$

respectively, for any $\theta \in [0, 2\pi)$. Moreover, it is known that $J(\theta) = HZ(\theta)$ generates any one-qubit gate [6,10]. A set of observables $\mathcal{S}$ is universal (resp. approximately universal) for quantum computation if there exists a universal (resp. approximately universal) set of gates such that any gate (that is, unitary operation) in the set can be simulated by using projective measurements described by the observables in $\mathcal{S}$.

3   Universal Quantum Computation

The previous simulation step of $J(\theta)$ is based on the state transfer [6], which uses X- and $Z \otimes Z$-measurements. For example, it implies a simulation step of H using Z-, X-, and $Z \otimes X$-measurements. We simplify this by using the state transfer based on Y-measurements depicted in Fig. 1. This implies a simulation step of a unitary operation using projective measurements depending on the operation. For example, we can obtain a simulation step of H by replacing $-Y$ and $Z \otimes Z$ with $H(-Y)H^{\dagger} = Y$ and $Z \otimes (HZH^{\dagger}) = Z \otimes X$, respectively. This uses only Y- and $Z \otimes X$-measurements and is simpler than the previous one.

On the basis of the idea, we show the following theorem:

Theorem 1. *The set of observables* $\mathcal{S}_1 = \{Z \otimes X, (\cos\theta)X + (\sin\theta)Y \text{ all } \theta \in [0, 2\pi)\}$ *with one ancillary qubit is universal for quantum computation.*

Proof. Since $J(\theta)$ generates any one-qubit gate and $\{\Lambda Z, J(\theta) \text{ all } \theta \in [0, 2\pi)\}$ is universal [6,10], the set of gates $\{(P^{-1} \otimes HP^{-1})\Lambda Z(I \otimes H), J(\theta) \text{ all } \theta \in [0, 2\pi)\}$ is universal, where $P = Z(\pi/2)$. Thus, to show the theorem, it suffices

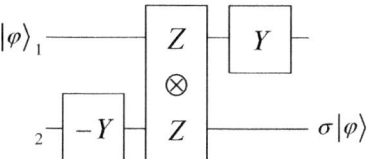

Fig. 1. The state transfer based on Y-measurements

to simulate any gate in the above set by projective measurements described by the observables in $\mathcal{S}_1$.

To give the simulation step of $J(\theta)$, we consider the procedure depicted in Fig. 2, which is obtained by replacing $-Y$, $Z \otimes Z$, and Y in Fig. 1 with $H(-Y)H^\dagger = Y$, $(Z(\theta)^\dagger ZZ(\theta)) \otimes (HZH^\dagger) = Z \otimes X$, and $Z(\theta)^\dagger YZ(\theta) = \cos(\pi/2 - \theta)X + \sin(\pi/2 - \theta)Y$, respectively. Let $|\varphi\rangle = \alpha|0\rangle + \beta|1\rangle$ and $s_1, s_2, s_3 \in \{1, -1\}$ be the classical outcomes of the measurements $Y^{(2)}$, $Z^{(1)} \otimes X^{(2)}$, and $(\cos(\pi/2 - \theta)X + \sin(\pi/2 - \theta)Y)^{(1)}$, respectively. The first measurement transforms the input state into

$$(I \otimes \sigma_z^{\frac{1-s_1}{2}})(\alpha|0\rangle + \beta|1\rangle)|+\tfrac{\pi}{2}\rangle.$$

The second measurement transforms the state into

$$(\sigma_z^{\frac{1-s_1 s_2}{2}} \otimes \sigma_z^{\frac{1-s_2}{2}})(\alpha|0\rangle|+\rangle - i\beta|1\rangle|-\rangle).$$

The third measurement transforms it into

$$(\sigma_z^{\frac{1-s_3}{2}} \otimes \sigma_z^{\frac{1-s_2}{2}} \sigma_x^{\frac{1+s_1 s_2 s_3}{2}})|+\tfrac{\pi}{2}-\theta\rangle(\alpha|+\rangle + e^{i\theta}\beta|-\rangle),$$

which is the desired output state since $J(\theta)|\varphi\rangle = \alpha|+\rangle + e^{i\theta}\beta|-\rangle$. Thus, the procedure depicted in Fig. 2 is a simulation step of $J(\theta)$, where $\sigma = I$, σ_x, σ_z, or $\sigma_z \sigma_x$ ($= \sigma_y$ up to a global phase). It can be shown that each σ occurs with the same probability, $1/4$.

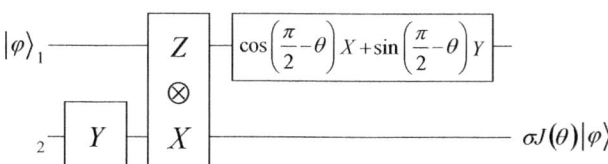

Fig. 2. The simulation step of $J(\theta)$

To give the simulation step of $(P^{-1} \otimes HP^{-1})\Lambda Z(I \otimes H)$, we consider the procedure depicted in Fig. 3. Let $|\varphi\rangle = \alpha|00\rangle + \beta|01\rangle + \gamma|10\rangle + \delta|11\rangle$ and $s_1, s_2, s_3, s_4 \in \{1, -1\}$ be the classical outcomes of the measurements $Y^{(3)}$ (the left one), $Z^{(1)} \otimes X^{(3)}$, $Z^{(3)} \otimes X^{(2)}$, and $Y^{(3)}$ (the right one), respectively. The first measurement transforms the input state into

$$(I \otimes I \otimes \sigma_z^{\frac{1-s_1}{2}})(\alpha|00\rangle + \beta|01\rangle + \gamma|10\rangle + \delta|11\rangle)|+\tfrac{\pi}{2}\rangle.$$

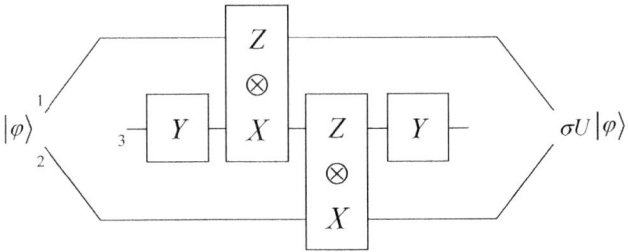

Fig. 3. The simulation step of $U = (P^{-1} \otimes HP^{-1})\Lambda Z(I \otimes H)$

The second measurement transforms the state into

$$(\sigma_z^{\frac{1-s_1s_2}{2}} \otimes I \otimes \sigma_z^{\frac{1-s_2}{2}})(\alpha|0\rangle|0\rangle|+\rangle + \beta|0\rangle|1\rangle|+\rangle - i\gamma|1\rangle|0\rangle|-\rangle - i\delta|1\rangle|1\rangle|-\rangle).$$

The third measurement transforms the state into

$$(\sigma_z^{\frac{1-s_1s_2s_3}{2}} \otimes \sigma_x^{\frac{1-s_2}{2}} \otimes \sigma_x^{\frac{1-s_3}{2}})(\alpha|0\rangle\frac{|+\rangle|0\rangle + |-\rangle|1\rangle}{\sqrt{2}} + \beta|0\rangle\frac{|+\rangle|0\rangle - |-\rangle|1\rangle}{\sqrt{2}}$$
$$-i\gamma|1\rangle\frac{|+\rangle|0\rangle - |-\rangle|1\rangle}{\sqrt{2}} - i\delta|1\rangle\frac{|+\rangle|0\rangle + |-\rangle|1\rangle}{\sqrt{2}}).$$

The fourth measurement transforms it into

$$(\sigma_z^{\frac{1-s_1s_2s_3}{2}} \otimes \sigma_x^{\frac{1-s_2s_3s_4}{2}} \otimes \sigma_z^{\frac{1-s_4}{2}})(\alpha|0\rangle\frac{|+\rangle - i|-\rangle}{\sqrt{2}} + \beta|0\rangle\frac{|+\rangle + i|-\rangle}{\sqrt{2}}$$
$$-i\gamma|1\rangle\frac{|+\rangle + i|-\rangle}{\sqrt{2}} - i\delta|1\rangle\frac{|+\rangle - i|-\rangle}{\sqrt{2}})|+\frac{\pi}{2}\rangle,$$

which is the desired output state. Thus, the procedure depicted in Fig. 3 is a simulation step of $(P^{-1} \otimes HP^{-1})\Lambda Z(I \otimes H)$, where $\sigma = I \otimes I$, $I \otimes \sigma_x$, $\sigma_z \otimes I$, or $\sigma_z \otimes \sigma_x$. It can be shown that each σ occurs with the same probability, $1/4$.

To implement σ_x and σ_z, we consider the procedures depicted in Figs. 4 and 5, respectively. It is easy to show that σ_x and σ_z are implemented with probability $1/2$. In the correction step, we repeat the procedure until the desired gate σ_x or σ_z is implemented as in [7]. The gate σ_y is implemented by combining the procedures. Thus, any gate in the set at the beginning of the proof can be simulated by projective measurements described by the observables in $\mathcal{S}_1$. □

The proof of Theorem 1 immediately implies Perdrix's result [7]:

Theorem 2. *The set of observables* $\mathcal{S}_2 = \{Z \otimes X, Y, (X + Y)/\sqrt{2}\}$ *with one ancillary qubit is approximately universal for quantum computation.*

Proof. Since $\{H, Z(\pi/4), \Lambda Z\}$ is approximately universal [9] and $(Z(\pi/4))^2 = P$, the set of gates $\{H, J(\pi/4), (P^{-1} \otimes HP^{-1})\Lambda Z(I \otimes H)\}$ is approximately universal. On the basis of the set of gates, it is easy to show the theorem. This is because, from the proof of Theorem 1, the simulation steps of $J(0) = H$ and $J(\pi/4)$ use projective measurements described by the observables only in $\mathcal{S}_2$. □

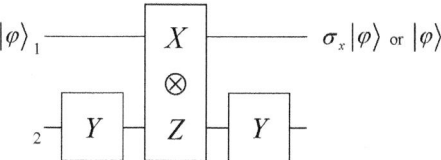

Fig. 4. The implementation of σ_x in the correction step

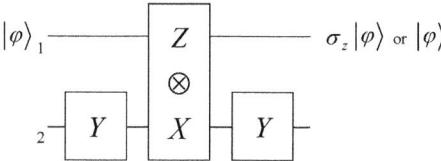

Fig. 5. The implementation of σ_z in the correction step

4 Graph State Preparation

Let $G = (V, E)$ be a graph with a set of vertices $V = \{1, \ldots, n\}$ and a set of edges $E \subseteq V \times V$. The corresponding graph state $|G\rangle$ is the quantum state obtained by the following procedure, where we assume that we have the initial state $|0\rangle_1 \cdots |0\rangle_n$ and call the k-th qubit the qubit corresponding to the vertex k:

1. Apply H to the qubit corresponding to the vertex k for any $k \in V$.
2. Apply ΛZ to the pair of qubits corresponding to the vertices k_1 and k_2 for any $(k_1, k_2) \in E$.

We call this procedure the standard procedure. For example, the graph state $|G\rangle$ corresponding to the graph G depicted in Fig. 6 is obtained by

$$\Lambda Z_{14} \Lambda Z_{23} \Lambda Z_{24} \Lambda Z_{34} H_1 H_2 H_3 H_4 |0\rangle_1 |0\rangle_2 |0\rangle_3 |0\rangle_4.$$

We consider a quantum circuit consisting of projective measurements. The complexity measures of a quantum circuit are the number of qubits in it and its size and depth as in the standard quantum circuit model. The size of a circuit is the number of measurements and the depth of a circuit is the number of layers in the circuit, where a layer consists of measurements that can be performed simultaneously. A quantum circuit can use ancillary qubits, which start in state $|0\rangle$.

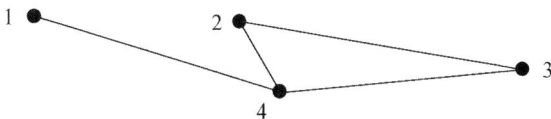

Fig. 6. The graph G with $V = \{1, 2, 3, 4\}$ and $E = \{(1, 4), (2, 3), (2, 4), (3, 4)\}$

We show that the set of observables $\mathcal{S}_3 = \{Z \otimes X, Y\}$ with one ancillary qubit is sufficient for preparing graph states efficiently. As described in the previous section, H can be simulated by using projective measurements described by the observables in $\mathcal{S}_3$. However, it is difficult to simulate P and thus ΛZ. Thus, it is difficult to use the standard procedure directly. Since we can simulate H and $(P^{-1} \otimes HP^{-1})\Lambda Z(I \otimes H)$, we can simulate $(P^{-1} \otimes P^{-1})\Lambda Z$.

Our circuit consists of three steps. In Step 2, we use $(P^{-1} \otimes P^{-1})\Lambda Z$ in place of ΛZ in Step 2 of the standard procedure. Since P^{-1} and ΛZ commute, this step is equivalent to Step 2 of the standard procedure up to local unitary gates generated by P^{-1}. We need to remove the side effects, that is, the local unitary gates, to obtain an exact graph state. If the degree of the vertex k is odd, the local unitary gate is P^{-1} or $(P^{-1})^3 = P$. If the degree is even, the local unitary gate is $(P^{-1})^2 = \sigma_z$ or $(P^{-1})^4 = I$.

Our idea of removing the side effects is that, if the degree of a vertex is odd, we perform a Y-measurement on the qubit corresponding to the vertex to prepare $P^{-1}H|0\rangle - |-\frac{\pi}{2}\rangle$ (or $PH|0\rangle - |+\frac{\pi}{2}\rangle$) in Step 1 of our circuit. If the degree is even, H (or $\sigma_z H$) is applied to the qubit. Combining Step 1 with Step 2 transforms the side effects in Step 2 to only σ_z or I. In Step 3, σ_z is removed if needed. For example, our circuit for preparing the graph state $|G\rangle$ corresponding to the graph G depicted in Fig. 6 is based on the circuit (in the standard quantum circuit model) depicted in Fig. 7.

Using the idea, we show the following theorem, where we assume that we have a classical description of a given graph and can thus use the degree of a vertex to construct a quantum circuit:

Theorem 3. *For any graph $G = (V, E)$, the graph state $|G\rangle$ can be prepared by a quantum circuit consisting of one-qubit and two-qubit projective measurements described by the observables in $\mathcal{S}_3 = \{Z \otimes X, Y\}$ with size and depth $O(n + m)$ and one ancillary qubit, where $n = |V|$ and $m = |E|$.*

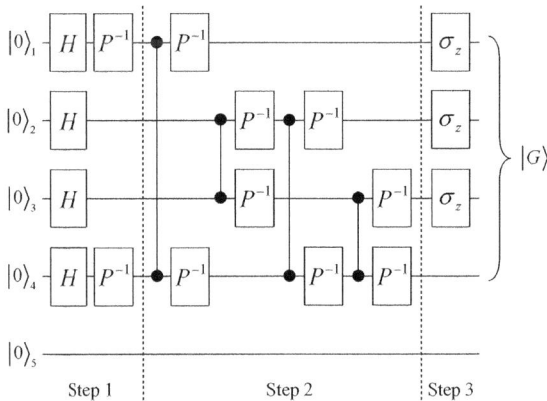

Fig. 7. The quantum circuit for preparing $|G\rangle$ corresponding to the graph G depicted in Fig. 6. The gate represented by two solid circles connected by a line is ΛZ. An ancillary qubit is reused to simulate $(P^{-1} \otimes P^{-1})\Lambda Z$ and σ_z.

Proof. Let $\deg(k)$ be the degree of the vertex k. We assume that we have the initial state $|0\rangle_1 \cdots |0\rangle_{n+1}$ and the $(n+1)$-th qubit is an ancillary qubit. Our circuit is constructed by using the following procedure:

1. For $k = 1, \ldots, n$:
   - If $\deg(k)$ is odd, perform a Y-measurement on the qubit corresponding to the vertex k. Let t_k be the classical outcome of the measurement.
   - If $\deg(k)$ is even, apply the simulation step of H where the k-th qubit is used as an ancillary qubit and the $(k+1)$-th qubit is used as an input qubit. Let u_k be the classical outcome of the $Z \otimes X$-measurement in the simulation step.
2. Apply $(P^{-1} \otimes P^{-1})\Lambda Z$ as in Step 2 of the standard procedure, where we reuse an ancillary qubit.
3. For $k = 1, \ldots, n$:
   If one of the following conditions holds, apply σ_z to the qubit corresponding to the vertex k, where we reuse an ancillary qubit:
   - $\deg(k) = 4a$ for some integer a and $u_k = -1$.
   - $\deg(k) = 4a + 1$ for some integer a and $t_k = -1$.
   - $\deg(k) = 4a + 2$ for some integer a and $u_k = 1$.
   - $\deg(k) = 4a + 3$ for some integer a and $t_k = 1$.

It is easy to show that the circuit works correctly and that the size and depth are $O(n + m)$ and the circuit uses only one ancillary qubit.   □

The depth is larger than Høyer et al.'s one. Since the graph states corresponding to connected graphs seem to be particularly useful in quantum information processing, we are interested in such graphs. For a connected graph, $m = \Omega(n)$ and thus the depth of our circuit is $O(m)$ in this case, which is asymptotically the same as Høyer et al.'s one.

5   Conclusions

We showed that the set of observables $\{Z \otimes X, (\cos \theta)X + (\sin \theta)Y$ all $\theta \in [0, 2\pi)\}$ with one ancillary qubit is universal. This improves Jorrand and Perdrix's result and the proof immediately implies Perdrix's result. The proof also implies that the set of observables $\{Z \otimes X, Y\}$ with one ancillary qubit is sufficient for preparing graph states efficiently. It would be interesting to investigate whether our result can be improved or not.

Acknowledgments. The author thanks Yasuhito Kawano, Seiichiro Tani, and Go Kato for their helpful comments.

References

1. Raussendorf, R., Briegel, H.J.: A One-Way Quantum Computer. Phys. Rev. Lett. 86, 5188–5191 (2001)
2. Nielsen, M.A.: Quantum Computation by Measurement and Quantum Memory. Phys. Lett. A 308, 96–100 (2003)

 3. Perdrix, S.: State Transfer Instead of Teleportation in Measurement-Based Quantum Computation. International Journal of Quantum Information 3(1), 219–223 (2005)
 4. Childs, A.M., Leung, D.W., Nielsen, M.A.: Unified Derivation of Measurement-Based Schemes for Quantum Computation. Phys. Rev. A 71, 032318 (2005)
 5. Leung, D.W.: Quantum Computation by Measurements. International Journal of Quantum Information 2(1), 33–43 (2004)
 6. Jorrand, P., Perdrix, S.: Unifying Quantum Computation with Projective Measurements Only and One-Way Quantum Computation. In: SPIE Quantum Informatics 2004, vol. 5833, pp. 44–51 (2005)
 7. Perdrix, S.: Towards Minimal Resources of Measurement-Based Quantum Computation. New Journal of Physics 9, 206 (2007)
 8. Høyer, P., Mhalla, M., Perdrix, S.: Resources Required for Preparing Graph States. In: Asano, T. (ed.) ISAAC 2006. LNCS, vol. 4288, pp. 638–649. Springer, Heidelberg (2006)
 9. Nielsen, M.A., Chuang, I.L.: Quantum Information and Quantum Computation. Cambridge University Press, Cambridge (2000)
10. Danos, V., Kashefi, E., Panangaden, P.: Parsimonious and Robust Realizations of Unitary Maps in the One-Way Model. Phys. Rev. A 72, 064301 (2005)

Computational Depth Complexity of Measurement-Based Quantum Computation

Dan Browne[1], Elham Kashefi[2], and Simon Perdrix[3]

[1] Department of Physics and Astronomy, University College London, UK
d.browne@ucl.ac.uk
[2] Laboratory for Foundations of Computer Science, University of Edinburgh, UK
ekashefi@inf.ed.ac.uk
[3] CNRS, Laboratoire d'Informatique de Grenoble, Grenoble University, France
simon.perdrix@imag.fr

Abstract. In this paper, we mainly prove that the "depth of computations" in the one-way model is equivalent, up to a classical side-processing of logarithmic depth, to the quantum circuit model augmented with unbounded fanout gates. It demonstrates that the one-way model is not only one of the most promising models of physical realisation, but also a very powerful model of quantum computation. It confirms and completes previous results which have pointed out, for some specific problems, a depth separation between the one-way model and the quantum circuit model. Since one-way model has the same parallel power as unbounded quantum fan-out circuits, the quantum Fourier transform can be approximated in constant depth in the one-way model, and thus the factorisation can be done by a polytime probabilistic classical algorithm which has access to a constant-depth one-way quantum computer. The extra power of the one-way model, comparing with the quantum circuit model, comes from its classical-quantum hybrid nature. We show that this extra power is reduced to the capability to perform unbounded classical parity gates in constant depth.

1 Introduction

The one-way quantum computational model, proposed by Raussendorf and Briegel [RB00], is remarkable in many aspects. It represents an approach to quantum computation very different to more conventional "circuit-based" approaches which were derived in close analogy to classical logic circuits. In the one-way model, computation proceeds by the generation of a particular entangled multi-qubit state - a cluster state - followed by the adaptive measurement of individual qubits. The choice of basis for the measurements, and their adaptive dependency encodes the computation.

The dependancy of the bases upon the outcome of previous measurements is a necessary part of the model. It compensates for the inherent randomness of the outcome of individual measurements allowing deterministic computation. Measurements which are not directly or indirectly dependent upon each other

W. van Dam et al. (Eds.): TQC 2010, LNCS 6519, pp. 35–46, 2011.

can be performed simultaneously. Thus the one-way model offers a radically different approach to the parallelisation of computations.

Broadbent and Kashefi [BK07] have pointed out a depth separation between the quantum circuit model and the one-way model. Indeed, there is a constant depth one-way quantum computation for implementing the parity gate, whereas there is no poly-size constant-depth circuit for this gate. Moreover they have proved that for any problem, the depth-separation between quantum circuits and one-way model is at most logarithmic factor. However, the exact parallel time power of the one-way quantum computation remained unknown.

In this paper, we mainly prove that the parallel time power of the one-way model is equivalent, up to classical side-processing, to the quantum circuit model augmented with unbounded fanout gates. This model, first explored by Høyer and Špalek [HS05] is a computational model which allows any two commuting gates to be performed simultaneously. The unbounded fanout model is surprisingly powerful, for example, the quantum component of Shor's algorithm reduces to constant depth in this model. Our results imply that the one-way model shares this power, provided the depth of the classical parity computations which make up the dependency calculations is neglected.

In section 2, we review the Measurement Calculus [DKP07, DKPP09], a formal framework in which the one-way model can be succinctly represented. Section 3 is dedicated to the comparison of the parallel time power of the one-way model and the quantum circuit model. In section 4 we present the unbounded fanout circuit model and prove the main result of this paper, its equivalence with the one-way model. In section 5, we show that reasonable classically controlled models can be efficiently implemented in the one-way model, answering an open question stated in [KOBAA09]. In section 6, we discuss some applications of these results and in section 7 we discuss the assumptions underlying each of these models.

2   Measurement Calculus

One-way quantum computations can be rigourously described in the measurement calculus formalism [DKP07]. A term of measurement calculus, called *measurement pattern* and playing the role of a circuit in the quantum circuit model, is a quadruplet (V, I, O, A). V is a finite register of qubits, $I, O \subseteq V$ are subregisters denoting respectively the inputs and output qubits. The non input qubits $(V \setminus I)$ are initialised in state $|+\rangle = \frac{1}{\sqrt{2}}(|0\rangle + |1\rangle)$. A is a sequence of quantum operations, called commands. They are three kinds of commands:

- $E_{i,j}$ is an entangling operation, which is nothing but the controlled-Z ΛZ unitary gate on qubits $i, j \in V$ where ΛZ is definied as follows:

$$\Lambda Z = \begin{pmatrix} 1 & 0 & 0 & 0 \\ 0 & 1 & 0 & 0 \\ 0 & 0 & 1 & 0 \\ 0 & 0 & 0 & -1 \end{pmatrix}$$

- M_i^α is a measurement of the qubit i in the basis $\{|+_\alpha\rangle, |-_\alpha\rangle\}$:

$$|\pm_\alpha\rangle = \frac{1}{\sqrt{2}}(|0\rangle \pm e^{i\alpha}|1\rangle) = \frac{1}{\sqrt{2}}\begin{pmatrix} 1 \\ \pm e^{i\alpha} \end{pmatrix}$$

This measurement produces a classical outcome $s_i \in \{0, 1\}$.

- X_i^s and Z_i^s are Pauli corrections or *dependant corrections*, where s is a finite sum modulo 2 of classical outcomes s_j's. For instance, $X_i^{s_j + \ldots + s_k}$ is applied on qubit i and depends on the sum of $s_j, \ldots, s_k$ as follows: $X^0 = I$ and $X^1 = X$.

$$X = \begin{pmatrix} 0 & 1 \\ 1 & 0 \end{pmatrix} \quad Z = \begin{pmatrix} 1 & 0 \\ 0 & -1 \end{pmatrix}$$

Measurements are supposed to be destructive (a measured qubit cannot be reused anymore), as a consequence a measurement pattern is well-formed if no command is applied on already measured qubits.

Example 1.

$$t = (\{1, 2, 3\}, \{1\}, \{3\}, X_3^{s_1 + s_2} M_2^\alpha X_2^{s_1} M_1^0 E_{1,2} E_{2,3} E_{1,3})$$

Note that the commands are read from right to left. This pattern is implementing an Z-rotation $R_z(-\alpha)$ from qubit 1 to qubit 3 (see section 3 for the definition of $R_z(\alpha)$).

Another kind of command, the *dependant measurements* exist in the measurement calculus. They can be defined as a combination of measurements and dependant corrections:

$$^\tau[M_i^\alpha]^\sigma := M_i^\alpha X_i^\sigma Z_i^\tau = M_i^{(-1)^\sigma \alpha + \tau\pi}$$

Corrections of the form $X^{s_j + \ldots + s_k}$ clearly illustrates the hybrid nature of measurement-based quantum computing: a classical control is collecting classical outcomes of previous measurements and computes the sum for deciding what the next quantum operation is.

Let $t_1 = (V_1, I_1, O_1, A_1)$ and $t_2 = (V_2, I_2, O_2, A_2)$ be two measurement patterns. The sequential composition $t_2 \circ t_1$ is defined as

$$t_2 \circ t_1 := (V_1 \cup V_2, I_1, O_2, A_2 A_1)$$

where we assume, up to a relabelling of the qubits in $V_2 \setminus O_1$ that $V_1 \cap V_2 = O_1 \cap I_2$. The parallel (or tensor) composition $t_1 \otimes t_2$ is defined as

$$t_1 \otimes t_2 := (V_1 \cup V_2, I_1 \cup I_2, O_1 \cup O_2, A_2 A_1)$$

where we assume, up to a relabelling of the qubits in V_2 that $V_1 \cap V_2 = \emptyset$.

The size of a command is the number of qubits affected by it. Notice that for any finite sums σ and τ of classical outcomes, X_i^σ, Z_i^τ and $^\tau[M_i^\alpha]^\sigma$ are of size 1. The *size* of a measurement pattern is the total size of all its commands. The classical dependency between one-qubit operations is an important ingredient of the depth of a measurement pattern. Indeed, a correction of the form $X_i^{s_j}$ has to be applied after the measurement of the qubit j.

The depth of a measurement pattern is the longest path of dependant commands:

Definition 1 (Quantum Depth). *For a given patterm* $t = (V, I, O, A)$, *its quantum depth* depth(t) *is defined as the longest sub-sequence* (p_x) *of* A *s.t. for any* x, $dom(p_x) \cap dom(p_{x+1}) \neq \emptyset$, *where* $dom(E_{i,j}) := \{i, j\}$, $dom(X_i^\sigma) = dom(Z_i^\sigma) := \{i\} \cup \{j$ *s.t.* s_j *appears in* $\sigma\}$, $dom(M_i^\alpha) := \{i\}$ *and* $dom(^\tau[M_i^\alpha]^\sigma) := \{i\} \cup \{j$ *s.t.* s_j *appears in* σ *or in* $\tau\}$.

Example 2. The pattern $t = (\{1, 2, 3\}, \{1\}, \{3\}, X_3^{s_1+s_2} M_2^\alpha X_2^{s_1} M_1^0 \, E_{2,3} E_{1,2} E_{1,3})$ is of size 10 and depth 6 ($X_3^{s_1+s_2} M_2^\alpha X_2^{s_1} M_1^0 E_{1,2} E_{1,3}$ is a dependant sub-sequence of size 6).

Notice that the quantum depth does not take into account the depth of the classical side-processing coming from computation of the classical sums. As a consequence, the quantum depth of a measurement pattern is based on the assumption that the classical computation is free, which can be motivated by the fact that the physical implementation of the quantum part of the computation is much more challenging than the classical part which can be considered at first approximation as free. This assumption is discussed in details in section 7.

For any measurement patterns t_1 and t_2, size$(t_1 \otimes t_2) =$ size$(t_2 \circ t_1) =$ size$(t_1) +$ size(t_2). Moreover, depth$(t_1 \otimes t_2) = max($depth$(t_1),$ depth$(t_2))$ and depth$(t_2 \circ t_1) \leq$ depth$(t_1) +$ depth(t_2).

Since $^\tau[M_i^\alpha]^\sigma = M_i^\alpha X_i^\sigma Z_i^\tau$, any measurement pattern t can be rewritten into a measurement pattern t' without dependant measurements such that size$(t') \leq 3.$size(t) and depth$(t') \leq 3.$depth(t). As a consequence, w.l.o.g, we consider in the rest of the paper only measurement patterns without dependant measurements.

In the following, we define some classes of complexity for measurement patterns. We consider only *uniform families*, whose description can be generated by a log-space Turing machine. Moreover we consider a fixed basis of measurement angles i.e. $0, \pi$ and an irrational multiple of π.

Definition 2. QMNC$(d(n))$ *contains decision problems computed exactly by uniform families of measurement patterns of input size* n, *depth* $O(d(n))$, *polynomial size, and over a fixed basis. Let* QMNC$^k =$ QMNC$(\log^k n)$ *and* BQMNC^k *contain decision problems computed with two-sided, polynomially small error.*

3   Measurement Patterns and Quantum Circuits

The comparison of the computational power of quantum circuits and measurement patterns has been extensively studied [RBB03, BK07]. Indeed, since the introduction of the one-way model, the advantage of the one-way model in terms of parallel time complexity has been pointed out on some examples. In this section, we review the main results and state them in terms of complexity classes.

A quantum circuit is a sequence of quantum gates. These gates are acting on three kinds of qubits: *input, output* and *ancilla* qubits. Input and output qubits may overlap[1]. A quantum circuit is a quadruplet $C = (V, I, O, G)$, where V is the set of all the qubits (input, output and ancilla qubits), $I, O \subset V$ are sets of input and output qubits. G is a sequence of gates. The size of a gate is the number of affected qubits, the size of a circuit is the total size of all its gates. The depth is longest path of dependant gates. A gate is a unitary operation applied on a bounded number of qubits. We consider the following one- and two-qubit gates: H, $R_z(\alpha)$ and ΛZ (see section 2 for a definition of ΛZ):

$$H := \frac{1}{\sqrt{2}} \begin{pmatrix} 1 & 1 \\ 1 & -1 \end{pmatrix} \quad R_z(\alpha) := \begin{pmatrix} 1 & 0 \\ 0 & e^{i\alpha} \end{pmatrix}$$

$\{H, R_z(\alpha_0), \Lambda Z\}$ with α_0 an irrational multiple of π is a universal family that can approximate any quantum gate with good precision [ADH97].

Running a quantum circuit consists in initialising the non input qubits in the $|0\rangle$ state, then applying sequence of gates, and finally measuring the non output qubits in the standard basis.

We consider the classes of complexity QNC and QNC^k for quantum circuits [MN98]: $\text{QNC}(d(n))$ contains decision problems computed exactly by uniform families of quantum circuits of input size n, depth $O(d(n))$, polynomial size, and over a fixed basis. Let $\text{QNC}^k = \text{QNC}(\log^k n)$, and BQNC^k contain decision problems computed with two-sided, polynomially small error.

Measurement patterns can be translated into quantum circuits and vice versa.

Lemma 1 ([RBB03]). *Any quantum circuit C can be simulated by a measurement pattern t of size $O(\text{size}(C))$ and depth $O(\text{depth}(C))$.*

Lemma 2 ([BK07], Lemma 7.9). *Any measurement pattern t can be simulated by a quantum circuit C of size $O(\text{size}(t)^3)$ and depth $O(\text{depth}(t) \log(\text{size}(t)))$.*

These lemmas lead to the following inclusions of complexity classes:

Theorem 1. *for any $k \in \mathbb{N}$,*

$$\text{QNC}^k \subseteq \text{QMNC}^k \subseteq \text{QNC}^{k+1}$$

$$\text{BQNC}^k \subseteq \text{BQMNC}^k \subseteq \text{BQNC}^{k+1}$$

Proof. Lemma 1 implies $\text{QNC}^k \subseteq \text{QMNC}^k$. Moreover, for any measurement pattern t of input size n and polynomial size, $\log(\text{size}(t)) = O(\log(n))$. So, according to Lemma 1, $\text{QMNC}^k \subseteq \text{QNC}^{k+1}$. $\qquad\square$

There is potentially a logarithmic depth separation between quantum circuits and measurement patterns, and indeed such a separation has been pointed out for the computation of the PARITY:

$$U^{(n)}_{\text{parity}} = |x_1, \ldots, x_n\rangle \mapsto |x_1, \ldots, x_{n-1}, \bigoplus_{i=1\ldots n} x_i\rangle$$

---

[1] This definition slightly generalises the usual definition of quantum circuit where input and output qubits are the same.

U_{parity} is a so called Clifford operation, a class of operations that can be computed in constant depth using measurements patterns [RBB03]. On the other hand, the depth of a quantum circuit for computing $U_{\text{parity}}^{(n)}$ is of depth $\Omega(\log(n))$ [FFGHZ03, BK07]. It implies that $\text{QNC}^0 \neq \text{QMNC}^0$. Such a separation is an open question for $k > 0$.

In the next section, we characterise the parallel time power of the measurement patterns using a reduction to quantum circuits with unbounded fan-out.

4 Measurement Patterns and Quantum Circuits with Unbounded Fan-Out

Measurement patterns and quantum circuits do not have the same computational power in terms of parallel time. In this section we compare the parallel power of the measurement patterns with a stronger version of the quantum circuits, the quantum circuits with unbounded fan-out. This model of quantum circuits was introduced by Høyer and Špalek [HS05]. The original motivation for introducing such a model of quantum circuits with unbounded fan-out circuits is that in addition to parallelising operations acting on distinct qubits, commuting operations can also, in certain circumstances, be applied simultaneously. See [HS05] for details on quantum fan-out circuits. A quantum circuit with unbounded fan-out is a circuit with the usual gates and also fan-out gates which have unbound input/output size:

Definition 3 (fan-out gate). *Fan-out gate maps*

$$U_{\text{fan-out}}^{(n)} = |y_1, \ldots, y_{n-1}, x\rangle \mapsto |y_1 \oplus x, \ldots, y_{n-1} \oplus x, x\rangle$$

The depth and size are defined like for quantum circuits. Notice that a fan-out gate acting on n qubits has size n. We consider the classes of complexity QNC_f and QNC_f^k for quantum circuits [HS05]: $\text{QNC}_f(d(n))$ contains decision problems computed exactly by uniform families of quantum circuits with unbounded fan-out of depth $O(d(n))$, polynomial size, and over a fixed basis. Let $\text{QNC}_f^k = \text{QNC}_f(\log^k n)$, and BQNC_f^k contain decision problems computed with two-sided, polynomially small error.

Lemma 3. *There exists a fan-out circuit of depth 3 which implements the parity gate*

$$U_{\text{parity}}^{(n)} = |x_1, \ldots, x_n\rangle \mapsto |x_1, \ldots, x_{n-1}, \bigoplus_{i=1 \ldots n} x_i\rangle$$

Proof. $U_{\text{parity}}^{(n)} = H^{\otimes n} \circ U_{\text{fan-out}}^{(n)} \circ H^{\otimes n}$ □

Quantum fan-out circuit can be used to parallelising commuting operations if a basis change making these operators diagonal can be implemented efficiently:

Lemma 4. *[HJ85, Theorem 1.3.19] For every set of pairwise commuting unitary gates, there exists an orthogonal basis in which all the gates are diagonal.*

Theorem 2. *[MN02, GHMP02] Let $\{U_i\}_{i=1}^n$ be pairwise commuting gates on k qubits. Gate U_i is controlled by qubit x_i. Let T be a gate changing the basis according to Lemma 4. There exists a quantum circuit with fan-out computing $U = \prod_{i=1}^n \Lambda_{x_i} U_i$ having depth $\max_{i=1}^n \mathrm{depth}(U_i) + 4 \cdot \mathrm{depth}(T) + 2$, size $\sum_{i=1}^n \mathrm{size}(U_i) + (2n+2) \cdot \mathrm{size}(T) + 2n$, and using $(n-1)k$ ancillas.*

In the following we prove that the quantum depth of the measurement patterns and the depth of the quantum circuits with unbounded fan-out coincide. First, we prove that measurement patterns are at least as powerful as quantum circuits with unbounded fan-out:

Lemma 5. *Any quantum circuit with unbounded fan-out C can be simulated by a measurement pattern of depth $O(\mathrm{depth}(C))$ and size $O(\mathrm{size}(C)^3)$.*

Proof. Each gate of C can be simulated by a measurement pattern. Indeed H, $R_z(\alpha_0)$, ΛZ can be simulated by constant depth, constant size measurement patterns:

$$t_H = (\{1,2\}, \{1\}, \{2\}, X_2^{s_1} M_1^0 E_{1,2})$$
$$t_{R_z(\alpha)} = t_H \circ (\{0,1\}, \{0\}, \{1\}, X_2^{s_1} M_1^\alpha E_{1,2})$$
$$t_{\Lambda Z} = (\{1,2\}, \{1,2\}, \{1,2\}, E_{1,2})$$

Since $U_{\mathrm{fan\text{-}out}}^{(n)}$ is equal to $H^{\otimes n} \circ U_{\mathrm{parity}}^{(n)} \circ H^{\otimes n}$ and since it exists a measurement pattern for $U_{\mathrm{parity}}^{(n)}$ of depth $O(1)$ and size $O(n^3)$ (see section 3), $U_{\mathrm{fan\text{-}out}}^{(n)}$ can be simulated by a measurement pattern of constant depth and $O(n^3)$ size. By sequential and parallel compositions, C is simulated by a measurement pattern t of size $O(\mathrm{size}(C)^3)$ and depth $O(\mathrm{depth}(C))$. □

Moreover, quantum circuits with unbounded fan out are at least as powerful as measurement patterns:

Lemma 6. *Any measurement pattern t can be implemented by a quantum circuit with unbounded fan-out of depth $O(\mathrm{depth}(t))$ and size $O(\mathrm{size}(t)^2)$.*

Proof. For a given measurement pattern $t = (V, I, O, A)$, the sequence of commands A can be rewritten into $k = \mathrm{depth}(t)$ layers $A^{(i)}$ of depth 1 such that $A = A^{(k)} \ldots A^{(1)}$. In the following we show that each of these layers can be translated into a constant depth piece of quantum fan-out circuit. Given a layer $A^{(i)}$, since $A^{(i)}$ is of depth 1, it implies that each operation is acting on distinct qubits. Thus, up to some commutations, we assume w.l.o.g. that the commands of $A^{(i)}$ are performed in the following order: the entangling operations first, followed by measurements, Z-corrections and finally X-corrections.

- The sub-sequence of $A^{(i)}$ composed of entangling operations is translated into a quantum circuit composed of ΛZ. The depth of this circuit is one since the entangling operations are acting on distinct qubits.
- The sub-sequence of $A^{(i)}$ composed of measurements is translated into a quantum circuit where each measurement $M_j^{\alpha_j}$ is replaced by a $H_j R_z(-\alpha_j)$. The depth of this circuit is 2.

- The sub-sequence of $A^{(i)}$ composed of Z-corrections is translated to a quantum circuit where each Z_j^σ is replaced by $\prod_{s_k \in \sigma} \Lambda_k Z_j$. This piece of circuit is not of constant depth, however, all the unitary transformations are diagonal, thus according to the Theorem 2, this piece of circuit can be simulated by quantum circuit with unbounded fan out of constant depth and size $O(\text{size}(t))$.
- Similarly, the sub-sequence of $A^{(i)}$ composed of X-corrections is translated into a quantum circuit with unbounded fan out of constant depth and size $O(\text{size}(t))$. In this case, since X is not diagonal, H is used as basis change in Theorem 2.

Thus each layer $A^{(i)}$ is translated to a piece of circuit of constant depth and size $O(s.\text{size}(t))$, where s is the size of $A^{(i)}$. As a consequence the whole pattern is translated into a quantum circuit with unbounded fan-out of detph $O(\text{depth}(t))$ and size $O(\text{size}(t)^2)$. $\qquad\square$

The combination of Lemmas 5 and 6 leads us to the main result of this paper. It implies that measurement patterns and quantum circuits with unbounded fan-out have the same computational power in terms of parallel time complexity:

Theorem 3. *For any $k \in \mathbb{N}$,*

$$\text{QMNC}^k = \text{QNC}_\text{f}^k$$

$$\text{BQMNC}^k = \text{BQNC}_\text{f}^k$$

5   Generalisation to Classically Controlled Quantum Computation

In this section a general scheme of classically controlled quantum computation is considered. One-way quantum computation is a special instance of this scheme as well as the teleportation based model [Nie03], the state transfer model [Per05] and the ancilla-driven quantum computation with twisted graph states [KOBAA09].

A classically controlled scheme, generalization of the measurement calculus, is characterized by a quadruplet $(\mathcal{I}, \mathcal{U}, \mathcal{O}, \mathcal{C})$ where $\mathcal{I}$ is a set of quantum states (for initialising ancillary qubits); $\mathcal{U}$ is a set of unitary transformations; $\mathcal{O}$ is a set of measurements with classical outcomes in $\{0, 1\}$; $\mathcal{C}$ is a set of corrections that are classically controlled by a sum modulo two of measurement outcomes.

For instance, the one-way model is a $(\{|+\rangle\}, \{\Lambda Z\}, \{\{|+_\alpha\rangle, |-_\alpha\rangle\}, \alpha \in [0, 2\pi)\}, \{Z, X\})$-scheme; the state transfer model [Per05] is a $(\emptyset, \emptyset, \{X \otimes Z, Z, (X - Y)/\sqrt{2}\}, \{Z, X\})$-scheme; the quantum circuit model is a $(\{|0\rangle\}, \{H, T, \Lambda Z\}, \{\{|0\rangle, |1\rangle\}\}, \emptyset)$-scheme; and the ancilla-driven model [KOBAA09] is a $(\{|+\rangle\}, \{(H \otimes H) \circ \Lambda Z\}, \{\{|+_\alpha\rangle, |-_\alpha\rangle\}, \alpha \in [0, 2\pi)\}, \{Z, X\})$-scheme.

Size and depth of a classically controlled pattern are defined as for the measurement patterns, where every primitive operation has a constant depth and its size is the number of qubits affected by it.

The classes QNC_f^k can be extended to unitary transformations, quantum states and quantum measurements as follows: a unitary U is in UQNC_f^k if U can be implemented by a quantum circuit with unbounded fan-out of depth k; a quantum state $|\phi\rangle$ is in IQNC_f^k if there exists $U \in \mathrm{UQNC}_f^k$ such that $|\phi\rangle = U|0\rangle$; finally a linear map O is in OQNC_f^k if there exists $U \in \mathrm{UQNC}_f^k$ such that $UOU^\dagger$ is diagonal. If $O \in \mathrm{OQNC}_f^k$ is an observable (self adjoint) then O is describing a measurement that can be implemented in depth $O(k)$ by a circuit with unbounded fan-out: the measurement according to O is transformed into a measurement in the standard basis thanks to the unitary $U \in \mathrm{OQNC}_f^k$ such that $UOU^\dagger$ is diagonal.

We show that among all the classically controlled schemes which commands can be implemented in constant depth, the measurement calculus is optimal in term of depth:

Theorem 4. *Given a classically controlled scheme* $(\mathcal{I}, \mathcal{U}, \mathcal{O}, \mathcal{C})$, *if* $\mathcal{I} \subseteq \mathrm{IQNC}_f^0$, $\mathcal{U} \subseteq \mathrm{UQNC}_f^0$, $\mathcal{O} \subseteq \mathrm{OQNC}_f^0$, $\mathcal{C} \subseteq \mathrm{UQNC}_f^0 \cap \mathrm{OQNC}_f^0$, *then any pattern* P *of that scheme can be implemented by a measurement pattern* t *of depth* $O(\mathrm{depth}(P))$.

Proof. We prove that any classically controlled pattern P can be implemented by a quantum circuit of unbounded fan-out of depth $O(\mathrm{depth}(P))$, which can then be implemented by a measurement pattern t of depth $O(\mathrm{depth}(P))$ according to lemma 6.

The proof generalizes the proof of lemma 6. P is composed of $\mathrm{depth}(P)$ layers $\{P^{(i)}\}_{1 \leq i \leq depth(P)}$. Notice that for each layer, the operations are acting on distinct qubits and then can be reorganised into subsequences of each type (intialisation, unitary, measurement and correction). Each of these subsequences can be implemented in constant depth:

- The subsequences of $P^{(i)}$ composed of initialisations, unitaries and measurments can be implemented in constant depth since $\mathcal{I} \subseteq \mathrm{IQNC}_f^0$, $\mathcal{U} \subseteq \mathrm{UQNC}_f^0$, and $\mathcal{O} \subseteq \mathrm{OQNC}_f^0$;
- The subsequence of $P^{(i)}$ composed of corrections can also be implemented in constant depth. Indeed, since every correction of the subsequence are acting on distinct qubit, all these corrections are commuting. Thus we can apply theorem 2, leading to a quantum circuit with unbounded fanout of constant depth since each $C \in \mathcal{C}$ is of constant depth and the gate T changing the basis is of constant depth as well since $\mathcal{C} \subseteq \mathrm{OQNC}_f^0$.

Thus each layer is implemented by a constant depth piece of circuit, so P is translated into a unbounded fan-out circuit of depth $\mathrm{depth}(P)$. $\square$

Notice that the one-way, the teleportation-based, the state-transfer-based, and the ancilla-driven models are all classically controlled models for which theorem 4 applies. As a corollary, it exists a depth-preserving translation from the ancilla-driven model to the one-way model, answering an open question stated in [KOBAA09].

6   Applications

The complexity classes of the quantum circuits with unbounded circuits have been studied and compared to other classes. Thanks to Theorem 3, all known complexity results about quantum circuits with unbounded fan-out can be applied to the measurement patterns. Among them, it is known that the quantum Fourier transform (QFT) is in $\mathrm{BQNC_f^0}$ [HS05], so QFT is in $\mathrm{BQMNC^0}$. Whereas the one-way model have mainly been introduced as a promising model of physical implementation, it turns out that this model is very powerful: QFT can be approximated in constant depth in the one-way model. This result confirms that the semi-classical quantum Fourier transform can be done efficiently [GN96].

Moreover, factorisation is in $\mathrm{RP[BQNC_f^0]} = \mathrm{RP[BQMNC^0]}$ [HS05], thus the factorisation can be approximated efficiently on a probabilistic machine which has access to a constant depth one-way quantum computer.

7   A Weaker Assumption

Parallelisation in the quantum circuit model is based on the following assumption: (i) gates acting on distinct qubits can be performed simultaneously. Measurement patterns and quantum circuits with unbounded fan-out have an extra parallel power compared to quantum circuits because they are based on stronger assumptions. The additional assumption for quantum circuits with unbounded fan-out is: (ii) commuting operations can be done simultaneously.[2] Whereas the additional assumption for measurement patterns is: (iii) classical part of any measurement pattern can be done in constant depth.

In this section, we investigate assumption (iii), and we compare assumptions (ii) and (iii) which lead to the same extra power. We show that these two assumptions can be related, and that assumption (iii) is weaker than (ii).

The classical part of a measurement pattern is reduced to the computation of sum modulo 2 (i.e., parity) [AB09] in the dependant correction commands of the form $X_i^{s_i+\ldots+s_j}$. Thus, assumption (iii) can be rephrased as: *boolean unbounded fan-in parity gates can be done in constant depth.*

Lemma 7. *If boolean unbounded fan-in parity gates can be done in constant depth, any measurement pattern t can be done in* $\mathrm{depth}(t)$ *quantum layers of constant-depth interspersed by constant depth classical layers.*

Notice that without assumption (iii), the classical parity gate on n bits can be computed in depth $O(\log(n))$. Thus,

Lemma 8. *Any measurement pattern t can be done in* $\mathrm{depth}(t)$ *constant-depth quantum layers interspersed by* $O(\log(\mathrm{size}(t)))$*-depth classical layers.*

Now we show that assumption (iii) is weaker than assumption (ii). Indeed, assumption (ii) is that commuting operations can be done simultaneously, which

---

[2] In fact assumption (ii) implies assumption (i).

implies that classical[3] commuting operations can be done simultaneously. So, the classical parity can be implemented in constant depth using classical control-Not which are commuting since target and controlled bits are not overlapping.

In other words, whereas the assumption associated to the quantum circuits with unbounded fan-out is that commuting operations can be done in parallel, the classical version of this assumption is enough for the measurement patterns.

8   Conclusion

In this paper, we have shown that the measurement patterns have an equivalent computational power in terms of parallel time to the quantum circuits enhanced with unbounded fanout. This characterises the power of the one-way model for parallelisation of algorithms and demonstrates that a number of quantum algorithms can be achieved in this model with a modest quantum depth. The equivalence of these two models is at first sight surprising. The parallelisation structure of the one-way model would appear to be radically different to the quantum circuit model. Nevertheless, our results indicate a close underlying connection.

There remain a number of open questions. Can the separation $QNC^0 \neq QMNC^0$ (proved by Broadbent and Kashefi with the parity problem) be proved for any k: $QNC^k \neq QMNC^k$? Or at least for $k = 1$. Moreover, given the examples above, to what extent can quantum depth be minimised for generic quantum algorithms? Recently, it has been shown that non-trivial quantum algorithms can be implemented with a single round of simultaneously applied commuting gates [SB08]. Moreover, in [J05] Richard Jozsa made the following conjecture:

"Any polynomial time quantum algorithm can be implemented with only $O(\log n)$ quantum layers interspersed with polynomial time classical computations. "

Our results may have an implication for fault tolerant quantum computation. Thresholds for fault tolerant computation are typically derived under the assumption of a polynomial computational depth. How could such fault tolerant models be relaxed if only constant or logarithmic quantum depth would suffice? We have focussed on classically controlled quantum computation where the classical control consists sums modulo two in our study. These results applies to one-way quantum computation and various measurement-based models, like the teleportation-based and the ancilla-driven models. But, other variants of measurement-based quantum computation with a different dependency structure [GE07] could have a quite different characterisation. We hope that this work motivates further study of this rich area.

References

[AB09]   Anders, J., Browne, D.E.: Computational Power of Correlations. Physical Review Letters 102, 050502 (2009)

[ADH97]   Adleman, L., DeMarrais, J., Huang, M.: Quantum computability. SIAM Journal on Computing 26, 1524–1540 (1997)

---

[3] U is classical if U maps basis states to basis states in the computational basis.

[BK07]   Broadbent, A., Kashefi, E.: Parallelizing quantum circuits. To appear
            in Theoretical Computer Science (2007) (arXiv.org preprint 0704.1736)
[DKP07]   Danos, V., Kashefi, E., Panangaden, P.: The measurement calculus. J.
            ACM 54(2) (2007)
[DKPP09]   Danos, V., Kashefi, E., Panangaden, P., Perdrix, S.: Semantic Tech-
            niques in quantum Computation. In: Extended measurement calculus.
            Cambridge University Press, Cambridge (2010)
[FFGHZ03]   Fang, M., Fenner, S., Green, F., Homer, S., Zhang, Y.: Quantum lower
            bounds for fanout. Quantum Information and Computation 6, 46–57
            (2003)
[GE07]   Gross, D., Eisert, J.: Novel schemes for measurement-based quantum
            computation. Physical Review Letters 98, 220503 (2007)
[GHMP02]   Green, F., Homer, S., Moore, C., Pollett, C.: Counting, fanout, and
            the complexity of quantum ACC. Quantum Information and Compu-
            tation 2(1), 35–65 (2002)
[GN96]   Griffiths, R.B., Niu, C.-s.: Semiclassical Fourier transform for quantum
            computation. Physical Review Letters 76, 3228–3231 (1996)
[HJ85]   Horn, R.A., Johnson, C.R.: Matrix Analysis. Cambridge University
            Press, Cambridge (1985)
[HS05]   Høyer, P., Špalek, R.: Quantum fan-out is powerful. Theory of Com-
            puting 1(1), 81–103 (2005)
[J05]   Jozsa, R.: An introduction to measurement based quantum computa-
            tion (2005), arXiv pre-print: quant-ph/0508124
[KOBAA09]   Kashefi, E., Oi, D.K.L., Browne, D., Anders, J., Andersson, E.: Twisted
            Graph States for Ancilla-driven Universal Quantum Computation.
            ENTCS 249, 307–331 (2009)
[MN98]   Moore, C., Nilsson, M.: Parallel Quantum Computation and Quantum
            Codes (1998), arXiv pre-print:quant-ph/9808027v1
[MN02]   Moore, C., Nilsson, M.: Parallel quantum computation and quantum
            codes. SIAM Journal on Computing 31(3), 799–815 (2002)
[Nie03]   Nielsen, M.A.: Universal quantum computation using only projective
            measurement, quantum memory, and preparation of the 0 state. Phys.
            Rev. A 308, 96–100 (2003)
[Per05]   Perdrix, S.: State transfer instead of teleportation in measurement-
            based quantum computation. International Journal of Quantum Infor-
            mation 3(1), 219–223 (2005)
[RB00]   Raussendorf, R., Briegel, H.J.: Quantum computing via measurements
            only. Physical Review Letters 86, 5188–5191 (2001)
[RBB03]   Raussendorf, R., Browne, D.E., Briegel, H.J.: Measurement-based
            quantum computation on cluster states. Physical Review A 68 (2003)
[SB08]   Shepherd, D., Bremner, M.J.: Instantaneous Quantum Computation
            (2008), arXiv pre-print:0809.0847v1

Local Equivalence of Surface Code States

Pradeep Sarvepalli and Robert Raussendorf

Department of Physics and Astronomy, University of British Columbia, Vancouver

Abstract. Surface code states are an important class of stabilizer states that play a prominent role in quantum information processing. In this paper we show that these states do not contain any counterexamples to the recently disproved LU-LC conjecture. In the process we show that surface codes do not have any encoded non-Clifford transversal gates. We also prove some interesting structural properties of the CSS surface code states. We show that these states can be characterized as a class of minor closed binary matroids. This characterization could be of independent interest in that it makes a connection with the theory of binary matroids.

1 Introduction

Stabilizer states are ubiquitous in quantum information theory—making their appearance in many diverse areas in quantum information processing ranging from quantum error correction, communication, cryptography, and (measurement based) computation. In view of their broad applications it is not surprising there are many questions related to stabilizer states. One particular question, that has attracted not a little attention, concerns the equivalence of stabilizer states.

Suppose we are given a pair of stabilizer states, then we can test the local Clifford equivalence of the states efficiently i.e. in polynomial time [15]. However, no such efficient algorithm is known for local unitary equivalence. In this context Rains [11] showed that for linear stabilizer codes, with distance greater than two, local unitary equivalence implies local Clifford equivalence. This implies that for stabilizer states in these codes, one could test local unitary equivalence efficiently. It was conjectured that two stabilizer states are local unitary equivalent if and only if they are local Clifford equivalent. This is the LU-LC conjecture, see for instance [14]. The results of Hein et al [6] who considered the local equivalence classes of stabilizer states up to seven qubits seemed to support this conjecture. The scales were tipped in favour of the conjecture due to the results of Van den Nest et al [16] who considered a generalization of Rains' results [11] and Zeng et al [17] who further strengthened the results of [16].

However, in a surprising development, it was shown by Ji et al [7] that the LU-LC conjecture is false. In view of the preceding discussion one might be tempted to interpret this result negatively. In a more positive light, it also reminds us how intricate a place Hilbert space is, often defying our preconceptions. As initial applications, stabilizer states disproving the LU-LC conjecture are useful resources

W. van Dam et al. (Eds.): TQC 2010, LNCS 6519, pp. 47–62, 2011.

in measurement-based quantum computation and are related to foundational aspects of quantum mechanics [12]—they imply proofs of the Kochen-Specker theorem [9].

Prompted by the results of [7] and the potential computational use of the counterexamples to the LU-LC conjecture, we ask if there are such counterexamples among prominent classes of stabilizer states. In this paper we focus on the surface code states. Our interest in them is partly borne out of their relevance for fault tolerant quantum computation, and partly to explore the possibility if there is any unifying structure, especially geometrical, underlying the counterexamples to the LU-LC conjecture. Hence, we study the local equivalence of the surface code states. We first consider *closed* surfaces of arbitrary genus. We show that under minimal restrictions on the states, they cannot be counterexamples to the LU-LC conjecture. We also extend this result to the case when the surface has boundaries but restrict our attention to the planar case.

In the process we prove a useful result, particularly relevant in the context of fault tolerant quantum computation. Quantum codes with encoded non-Clifford transversal gates are somewhat hard to find and desirable for fault tolerant quantum computation. Surface codes [8] are of great interest because they are especially suited for fault tolerant computation and their potential for high thresholds. However, no non-Clifford transversal gates have been found for these codes since their discovery almost a decade ago. As a consequence of our results we show that surface codes (on closed surfaces) do not have any non-Clifford transversal gates.

In addition we investigate some mathematical structures associated with the Calderbank-Shor-Steane (CSS) surface code states. These states can be characterized in terms of a class of minor closed matroids, in that they can be compactly described by finite list of matroids. This characterization could be potentially useful in its own right. An expanded version of this paper exploring local equivalence of stabilizer states in more detail is available at [13].

2   Background

We assume that the reader is familiar with the stabilizer formalism and quantum codes. We denote by $U(2^n)$ the group of $2^n \times 2^n$ unitary matrices. An element of $U(2^n)$ is local unitary if it is in the local unitary group $\mathcal{U}_n^l = U(2)^{\otimes^n}$. Two stabilizer states $|\psi\rangle$ and $|\psi'\rangle$ are said to be local unitary (LU) equivalent if $|\psi'\rangle = U|\psi\rangle$ for some $U \in \mathcal{U}_n^l$. We denote the local unitary equivalence class of a stabilizer state $|\psi\rangle$ by $\text{LU}(\psi)$.

The Clifford group over n qubits is the normalizer of $\mathcal{P}_n$, the Pauli group, in $U(2^n)$. In other words,

$$\mathcal{K}_n = \{U \in U(2^n) \mid U\mathcal{P}_n U^\dagger = \mathcal{P}_n\}. \tag{1}$$

The local Clifford group over n qubits is defined as $\mathcal{K}_n^l = \mathcal{K}_1^{\otimes^n}$. Two stabilizer states $|\psi\rangle$ and $|\psi'\rangle$ are local Clifford (LC) equivalent if there exists a $K \in \mathcal{K}_n^l$ such that $|\psi'\rangle = K|\psi\rangle$. We denote the local Clifford equivalence class of $|\psi\rangle$ by $\text{LC}(\psi)$.

An important tool to study the local equivalence of stabilizer states is the notion of the so-called minimal support elements. The support of an element $g = \otimes_{i=1}^{n} g_i$ in $\mathcal{P}_n$ is defined as the subset of $\{1, \ldots, n\}$ for which $g_i \neq I$ i.e.

$$\mathrm{supp}(g) = \{i \mid g_i \neq I\}. \tag{2}$$

Given a stabilizer state $|\psi\rangle$ we denote its stabilizer by $S(\psi)$. We say that $g \in S(\psi)$ is a minimal support element if there does not exist any element $h \in S(\psi)$ such that $\emptyset \neq \mathrm{supp}(h) \subset \mathrm{supp}(g)$. In other words, $\mathrm{supp}(g)$ does not strictly contain the support of any nontrivial element of S. We also say that g is a minimal element of $S(\psi)$, and $\mathrm{supp}(g)$ is a minimal support of $S(\psi)$. We define the weight of $g \in \mathcal{P}_n$ as

$$\mathrm{wt}(g) = |\{i \mid g_i \neq I\}| = |\mathrm{supp}(g)|. \tag{3}$$

The distance of a subgroup S in $\mathcal{P}_n$ is defined as $\min_{g \in S \setminus I} \mathrm{wt}(g)$, where g is not the identity element of S. Often, we refer to the distance of a stabilizer state by which we mean the distance of its stabilizer $S(\psi)$.

The action of local unitaries on the stabilizer provides an important handle in understanding the LU-LC equivalence classes of that state. In particular, the action of local unitaries on the minimal supports of a stabilizer state is of great significance. We state the relevant result below and refer the interested reader to [11] for further details.

Lemma 1. *If $U \in \mathcal{U}_n^l$ maps a stabilizer state $|\psi\rangle$ to another stabilizer state $|\psi'\rangle = U|\psi\rangle$, then U maps the minimal support elements of $S(\psi)$ to the minimal support elements of $S(\psi')$.*

If both the local equivalence classes of $|\psi\rangle$ are the same, then we indicate this by $\mathrm{LU}(\psi) = \mathrm{LC}(\psi)$. There are some conditions under which we can conclude that $\mathrm{LU}(\psi) = \mathrm{LC}(\psi)$. A sufficient condition due to van den Nest et al will be useful in this context. For proof and further details, please refer to [16, Theorem 1] and [16, Corollary 1].

Lemma 2 (Minimal Support Condition). *Suppose that $|\psi\rangle$ is a n-qubit stabilizer state free from Bell pairs. Let its stabilizer be $S(\psi)$ and $M(\psi)$ be the group generated by the minimal support elements of $S(\psi)$. If all the Pauli matrices X, Y, Z occur on every qubit in $M(\psi)$, (i.e. for any $\alpha \in \{X, Y, Z\}$, and for any i, there exists some $\otimes_{j=1}^{n} g_j \in M(\psi)$ such that it $g_i = \alpha$), then $\mathrm{LU}(\psi) = \mathrm{LC}(\psi)$. In particular this condition holds if $S(\psi) = M(\psi)$.*

Remark 1. We note that Lemma 2 slightly extends the result in [16, Theorem 1], in that instead of fully entangled stabilizer states we allow for stabilizer states that are free from Bell pairs. It can be shown that the original proof holds with this modified condition.

2.1   Some Notions from Graph Theory

Let Γ be a graph with vertex set $V(\Gamma)$ and edge set $E(\Gamma)$. A cycle is an alternating sequence of vertices and edges such that every edge connects the adjacent vertices and only the first and last vertices are same. The number of edges in a cycle is called the length of the cycle and a cycle of length n is called an n-cycle. A loop or 1-cycle is an edge connecting a vertex to itself.

It is not possible to draw every graph on a sphere so that its edges do not cross. But by adding handles to the sphere we can draw the graph such that its edges do not cross. The genus of a graph is the minimum number of handles that we need to add to the sphere so that it can be drawn without edge crossings. We call this an embedding of the graph. Let the genus of the graph be g and assume that it is embedded on a surface of genus $g' \geq g$. We denote the set of faces of Γ by $F(\Gamma)$. The union of all the faces equals the surface on which the graph is embedded.

Another notion that we need is that of dual graphs. Given a graph Γ, that is embedded on a surface, we can define a dual graph, denoted Γ^*. To form the dual graph Γ^*, we replace every face f with a vertex f^*, and connect two vertices f_1^* and f_2^* if f_1 and f_2 share an edge. It can be checked that the edges in Γ are in 1-1 correspondence with the edges in Γ^* while the vertices in Γ correspond to the faces in Γ^*. The edges incident on a vertex in Γ (Γ^*) are precisely the edges that form the boundary of the associated face in Γ^* (Γ). So the operator A_v is associated with a vertex and its incident edges in Γ, but it is associated with a face and its boundary edges in Γ^*. A set of edges is called an elementary cycle if it forms the boundary of a face in Γ. A set of edges of Γ which form a cycle in Γ^* is called a cocycle of Γ. A cocycle is called an elementary cocycle if it forms the boundary of a face in Γ^*. A loop is an edge connected to the same vertex. A coloop is an edge that forms a loop in the dual graph.

A connected graph is said to be 2-connected if it remains connected after removing any vertex and the edges incident on that vertex. Later when we study planar surface codes we need the following result about 2-connected graphs; see [2, Proposition 4.2.6] for a proof.

Lemma 3. *In a 2-connected plane graph, every face is bounded by a cycle.*

We also need the following notions on graph minors in Section 4. The graph obtained from Γ by deleting an edge e is denoted as $\Gamma \setminus e$ while the graph obtained by edge contraction is denoted as Γ/e. A graph obtained from Γ by a sequence of edge deletions and contractions is called a (graph) minor of Γ. The following result on graph minors is well known.

Lemma 4. *Supposing we have a graph Γ and its dual Γ^* and $e \in E(\Gamma)$. Then $(\Gamma \setminus e)^* = \Gamma^*/e$ and $(\Gamma/e)^* = \Gamma^* \setminus e$.*

2.2   Surface Code States

Surface codes were introduced by Kitaev [8] with a view to perform fault tolerant quantum computation. These codes have interesting connections with graphs and many of their properties can be analyzed in graph theoretic terms. In this section we provide a brief introduction to these codes and the stabilizer states that arise from these codes. Kitaev showed [8] we can associate a stabilizer code with Γ.

We identify qubits with the edges of the graph. Define the site operators and the face operators of Γ as:

$$A_v = \prod_{e \in \delta(v)} X_e; \quad B_f = \prod_{e \in \partial(f)} Z_e, \tag{4}$$

where $\delta(v)$ is the set of edges incident on the vertex v and $\partial(f)$ is the set of edges that constitute the boundary of the face f. Let $S = \langle A_v, B_f \mid v \in V(\Gamma), f \in F(\Gamma) \rangle$, i.e, the group generated by the site operators and the face operators. The code stabilized by S is said to be the surface code of Γ; a stabilizer state in the surface code is called a surface code state and denote it as $|\psi_\Gamma\rangle$.

The parameters of the surface codes can be obtained from the graph and its embedding. If Γ has no loops and has n edges and is embedded on a surface of genus g, the group generated by the site operators and the face operators contains $n-2g$ generators and it defines an $[[n, 2g]]$ surface code. In this case, the supports of the encoded operators are defined by the nontrivial cycles and cocycles of Γ. When the graph has loops or when its genus is smaller than the genus of the surface on which it has been embedded, the number of encoded qubits can vary. For a graph with n edges and genus g that is embedded on a surface of genus $g' \geq g$, the stabilizer has $n - k$ generators and $2k$ encoded operators, where $k \leq 2g'$. Denote the encoded operators as $\mathcal{L} = \{\overline{X}_1, \overline{Z}_1, \ldots, \overline{X}_k, \overline{Z}_k\}$. The support of any encoded operators is a nontrivial cycle of the surface. These encoded operators allow us to specify each surface code state as follows. Let $\mathfrak{C}_1(\Gamma)$ and $\mathfrak{C}_1(\Gamma^*)$ be the set of homologically nontrivial cycles of Γ and Γ^* respectively. Then $\mathrm{supp}(\overline{X}_i) \in \mathfrak{C}_1(\Gamma^*)$ while $\mathrm{supp}(\overline{Z}_j) \in \mathfrak{C}_1(\Gamma)$. A CSS surface code state is stabilized by

$$S = \left\langle A_v, B_f, \overline{X}_1, \ldots, \overline{X}_l, \overline{Z}_{l+1}, \ldots, \overline{Z}_k \,\middle|\, \begin{matrix} v \in V(\Gamma) \\ f \in F(\Gamma) \end{matrix} \right\rangle, \tag{5}$$

where we renumber the $\overline{X}_i$ and $\overline{Z}_j$ if necessary.

We need the following lemma (previously known) in the context of surface codes.

Lemma 5. *Let $|\psi\rangle$ be a surface code state associated to a graph Γ without loops or coloops. Let g be an element in $C(S(\psi))$, the centralizer of $S(\psi)$ such that g consists of only Z or X operators alone. Then $\mathrm{supp}(g)$ is a union of cycles (or cocycles) in Γ. In particular, the face (vertex) operators of Γ generate only union of cycles (cocycles) of Γ.*

Later on in Section 3 we will also consider surfaces with boundaries. However for ease of presentation, we will postpone the details of these surface codes with boundaries to a later part of Section 3.

3   LU-LC Equivalence of Surface Code States

In the first part of this section we study the local equivalence of stabilizer states on closed surfaces. In the second part we extend these results to stabilizer states on planar surfaces with boundaries. The reader must bear this distinction in mind to avoid potential confusion as the results, although similar, vary in the assumptions about the associated graphs, which require us to engage in a separate discussion. Then in the third part of the section we give an application of these results for fault tolerant quantum computing.

The central result is that surface code states cannot be counterexamples to the LU-LC conjecture. These results prepare the way to make the connection to matroids. Throughout this section we assume that the graphs are connected and that they free from cycles of length ≤ 2; thus the associated stabilizer states are free from Bell pairs. Some minor additional restrictions are imposed for planar surface code states.

In the following we often denote the stabilizer of a quantum code Q by $S(Q)$. Given a subset $\omega \subseteq \{1, 2, \ldots, n\}$, and an abelian subgroup $S \leq C(S(Q))$ we denote by $A_\omega^S = |\{g \in S : \text{supp}(g) = \omega\}|$.

3.1   Codes on Closed Surfaces

In this part of the section we focus on stabilizer states derived from surfaces codes on closed surfaces of arbitrary genus. The next lemma shows that every site operator A_v (face operator B_f) are either minimal or can be written as the product of minimal X-only (Z-only) operators.

Lemma 6. *Suppose Q is a surface code associated with a graph Γ without cycles or cocycles of length ≤ 2. Let S be an abelian subgroup of $C(S(Q))$ that contains $S(Q)$. Then for any $A_v \in S$ and $B_f \in S$ there exist minimal elements $\{g_1^v, \ldots, g_{k_v}^v\} \subset S$ and $\{h_1^f, \ldots, h_{k_f}^f\} \subset S$ such that*

i) Each g_i^v is an X-only operator and $\text{supp}(A_v) = \cup_{i=1}^{k_v} \text{supp}(g_i^v)$.

ii) $A_{\text{supp}(g_i^v)}^S = 1$.

iii) Each h_i^f is a Z-only operator and $\text{supp}(B_f) = \cup_{i=1}^{k_f} \text{supp}(h_i^f)$.

iv) $A_{\text{supp}(h_i^f)}^S = 1$.

Further, for any $j \in \{1, 2, \ldots, n\}$, there exists some A_v and B_f such that $j \in \text{supp}(A_v) \cap \text{supp}(B_f)$, where $n = |E(\Gamma)|$, the total number of qubits.

Proof. If A_v is minimal then i) holds trivially for A_v by letting $g_1^v = A_v$. So does ii) because in this case $A_{\text{supp}(A_v)}^S \neq 1$ implies that Γ has 2-cycles. Suppose that A_v is a non-minimal element in S. There exists a minimal element $g_1^v \neq I$ in S such that $\text{supp}(g_1^v) \subsetneq \text{supp}(A_v)$. Now as Q is a CSS code, g_1^v can be written as $g_1^v = g_x g_z$, for some $g_x, g_z \in C(S(Q))$ where both g_x and g_z are not simultaneously trivial. It follows that the supports of both g_x and g_z must be strictly contained in the support of A_v. In particular, $\text{supp}(g_z) \subsetneq \text{supp}(A_v)$. By Lemma 5, g_z must be a cycle of Γ. However, all the edges of g_z being a subset of $\text{supp}(A_v)$ must be incident on the same vertex. Then $\text{supp}(g_z)$ can be a cycle only if it contains 2-cycles. But this contradicts that Γ does not have 2-cycles. Therefore, $g_z = I$ and $g_x = g_1^v \neq I$. Now consider the element $g' = A_v g_1^v$, it is an X-only operator whose support is given by $\text{supp}(A_v) \setminus \text{supp}(g_1^v)$. Now g' is either minimal or not. If it is minimal then by relabeling $g' = g_2^v$ we can write $\mathrm{supp}(A_v) = \mathrm{supp}(g_1^v) \cup \mathrm{supp}(g_2^v)$ and we are done. On the other hand if g' is not minimal, we can repeat the same process as with A_v and we will eventually end up with a set of elements $\{g_1^v, \ldots, g_{k_v}^v\}$ such that $\mathrm{supp}(A_v) = \cup_{i=1}^{k_v}\mathrm{supp}(g_i^v)$. This shows that i) holds.

Let $\omega = \mathrm{supp}(g_i^v)$. Suppose that $A_\omega^S \neq 1$. Then there exists a $g_i' \in S$ such that $g_i' \neq g_i$ and $\mathrm{supp}(g_i') = \omega$. Since $\omega \subsetneq \mathrm{supp}(A_v)$, it follows that g_i' is also a minimal element within the support of A_v. But we have already seen that every minimal element in the support of A_v must consist of X-only operators. Therefore $g_i' = g_i^v$ contradicting that $g_i' \neq g_i^v$. Thus $A_{\mathrm{supp}(g_i^v)}^S = 1$, proving ii).

We claim that $j \in \{1, 2, \ldots, n\}$ occurs in the support of A_v for some $v \in V(\Gamma)$. Since an edge is incident on at most two vertices, if $j \notin \mathrm{supp}(A_v)$ for any $v \in V(\Gamma)$, this means that both ends of the edge associated with j must be incident on the same vertex. This implies that Γ has loops contrary to our assumptions. So every j occurs in the support of some site operator A_v.

By a similar argument but working in the dual graph Γ^*, we can show that $\mathrm{supp}(B_f) = \cup_{i=1}^{k_f}\mathrm{supp}(h_i)$ for some Z-only minimal elements in S and that j occurs in the support of some face operator B_f as long as there are no coloops.

Theorem 1. *Let Q be the surface code associated with a graph Γ cycles or cocycles of length ≤ 2. Let $S(Q)$ be the stabilizer of Q and $C(S(Q))$, the centralizer of $S(Q)$. Then for any surface code (stabilizer) state $|\psi\rangle$ we have $\mathrm{LU}(\psi) = \mathrm{LC}(\psi)$.*

Proof. The stabilizer of a general surface code state is a subgroup of $C(S(Q))$ and contains $S(Q)$.

$$S(\psi) = \left\langle S, L_1, L_2, \ldots, L_k \,\middle|\, \begin{array}{l} L_i \in C(S(Q)) \setminus SZ(S(Q)) \\ \text{and } L_iL_j = L_jL_i \end{array} \right\rangle,$$

where k is such that $S(\psi)$ is the stabilizer of $|\psi\rangle$. Let $n = |E(\Gamma)|$, the total number of qubits. From Lemma 6, we know that every vertex operator and face operator are such that there exist X-only minimal elements $\{g_1, \ldots, g_{k_v}\} \in S(\psi)$ and Z-only minimal elements $\{h_1, \ldots, h_{k_f}\} \in S(\psi)$ such that $\cup_{i=1}^{k_v}\mathrm{supp}(g_i) = \mathrm{supp}(A_v)$ and $\mathrm{supp}(B_f) = \cup_{i=1}^{k_f}\mathrm{supp}(h_i)$. Since for every j we have $j \in \mathrm{supp}(A_v) \cap \mathrm{supp}(B_f)$ for some $v \in V(\Gamma)$ and $f \in F(\Gamma)$, there exist minimal elements g, h such that $g_j = X$ and $h_j = Z$. This means that X, Y, Z occur on every qubit in $M(\psi)$. Since there are no 2-cycles or 2-cocycles, $S(\psi)$ cannot contain elements of the form X_iX_j or Z_iZ_j. Since Q is a CSS code if X_iZ_j or X_iY_j are in $C(S(Q))$, then it follows that Γ has loops or coloops, contradicting the assumptions on Γ. Thus $|\psi\rangle$ is free of Bell pairs. By Lemma 2, it follows that $\mathrm{LU}(\psi) = \mathrm{LC}(\psi)$.

3.2   Codes on Planar Surfaces with Boundaries

Surface codes take us closer to performing fault tolerant quantum computation with realistic physical systems. Nonetheless, one could object that having to work with surfaces of nontrivial genus could still pose a challenging problem practically. This becomes all the more relevant since surface codes defined on a plane can encode only one state, in order to encode at least one qubit we have to use surfaces of nontrivial genus. Fortunately, Bravyi and Kitaev [1] and independently Freedman and Meyer [4] proposed surface codes on planar surfaces which can encode more than one state. Their approaches while equivalent are in some sense very different. The approach taken in [1] makes use of the ideas of rough and smooth "boundaries" while that of [4] uses the notion of "punctured discs" or equivalently, "holes". Perhaps, the most appropriate language to study these codes in either approach is that of relative homology. Fortunately, for our purposes we can prove useful results without using the heavy artillery of relative homology.

We make use of approach developed by Freedman and Meyer [4] to define surface codes on plane. The essential idea of [4] is to promote some of the face operators to encoded operators, while retaining all the vertex operators in the stabilizer. In other words, given the face operators and site operators of a planar graph, some of the face operators are excluded from the stabilizer of the code. The main requirement on these faces is that no two faces share an edge in common.

Let Γ be a planar graph. When we embed a graph (of genus zero) on a plane, one of the faces is unbounded (in area); this unbounded or infinite face is also called the outer or exterior face. Given this planar embedding of Γ we associate with each face and each vertex an operator as defined earlier in equation (4). We denote the exterior face by f_∞ and the face operator associated with it by A_∞. There are $|F(\Gamma)|$ such operators including the A_∞ of which $|F(\Gamma)| - 1$ are independent. There are $|V(\Gamma)|$ vertex operators of which $|V(\Gamma)| - 1$ are independent.

In order to define a surface code on a plane that can encode $k > 0$ qubits, we first pick a set of $k + 1$ faces $H \subsetneq F(\Gamma)$ such that the following conditions hold:

$$f_\infty \in H, \text{ and } \partial(f_i) \cap \partial(f_j) = \emptyset \text{ for any pair of faces } f_i, f_j \in H. \qquad (6)$$

In other words, the external face is in H and no pair of faces in H share an edge in common. We call the faces in $H \setminus f_\infty$ the "punctured discs" or "holes" of Γ. The stabilizer of the corresponding planar surface code Q is defined as

$$S(Q) = \left\langle A_v, B_f \middle| \begin{matrix} v \in V(\Gamma) \\ f \in F(\Gamma) \setminus H \end{matrix} \right\rangle. \qquad (7)$$

We can compute the parameters of this surface code easily as follows. Excluding the k face operators and the external face, we have $|F(\Gamma)| - k - 1$, independent face operators in the stabilizer of the code. Together with the $|V(\Gamma)| - 1$ independent vertex operators, this implies that the encoded qubits are $|E(\Gamma)| - (|F(\Gamma)| - k - 1 + |V(\Gamma)| - 1) = k$, where we used the fact that $\chi = |V(\Gamma)| - |E(\Gamma)| + |F(\Gamma)| = 2$, is the Euler characteristic. Note that $|F(\Gamma)|$ includes the exterior face. From this it follows that the number of encoded qubits is k.

A canonical set of encoded operators for the planar surface code are given as follows. The encoded Z operators of the planar surface code are given by the face operators B_f, where $f \in H \setminus f_\infty$. The encoded X operators are a little more subtle to define. Suppose that each face f corresponds to the vertex f^* in the dual graph Γ^* and the exterior face corresponds to the vertex f_∞^*. We define a path in Γ to be a nonrepeating sequence of vertices $\{v_0, v_1, \ldots, v_l\}$ containing the edges $\{(v_0, v_1), \ldots, (v_{l-1}, v_l)\}$. Consider an excluded face $f_j \in H$. Let P be a path in Γ^* connecting the vertex f_∞^* and f_j^*. The encoded X operator $\overline{X}_j$ is given by

$$\overline{X}_j = \prod_{e \in P} X_e, \tag{8}$$

where e is an edge in P. In order to satisfy the commutation relations for the encoded operators, we choose that these paths intersect only at f_∞^*. The encoded X operator $\overline{X}_j$ anticommutes with the encoded Z operator B_{f_j}. The centralizer of $S(Q)$ is given by

$$C(S(Q)) = \langle A_v, B_f, \overline{X}_j, iI \,|\, v \in V(\Gamma); f \in F(\Gamma); 1 \le j \le k \rangle. \tag{9}$$

We have seen in Lemma 5, that the supports of encoded X (Z) operators of the surface codes are unions of cocycles (cycles) in Γ. However, in case of planar surface codes the analogous result is a little more complex.

Lemma 7. *Let Q be a planar surface code derived from a planar graph Γ without loops or coloops. We assume that the holes of Γ are defined by the set of faces $H \subsetneq F(\Gamma)$ satisfying equation (6). Let H^* denote the set of vertices in Γ^* corresponding to the faces in H. Suppose that $g \in C(S(Q))$. If g is a Z-only operator, then $\mathrm{supp}(g)$ is a union of cycles in Γ. If g is an X-only operator, then $\mathrm{supp}(g)$ is a union of cycles and/or paths in Γ^*. The paths must begin and end on vertices in H^*.*

Proof. Suppose that g is a Z-only operator in $C(S(Q))$ and $\mathrm{supp}(g)$ is not a union of cycles in Γ. Since g consists of only Z operators it must be generated by the face operators of Γ and the encoded Z operators of the surface code associated with Γ. Because $\mathrm{supp}(g)$ is not a union of cycles there exists a vertex $v \in V(\Gamma)$ such that an odd number of the edges of $\mathrm{supp}(g)$ are incident on v. Then g cannot commute with the vertex operator A_v (consisting of X operators alone) because A_v overlaps with g over odd number of qubits. Thus g cannot be an element of $C(S(Q))$, giving us a contradiction. Hence $\mathrm{supp}(g)$ must be a union of cycles.

Assume now that g is an X-only operator in $C(S(Q))$. Clearly, the support of g is a union of cycles and paths in Γ^*. *What we need to show is that all of the paths begin and end in vertices in H^*.* Suppose that $\mathrm{supp}(g)$ contains a path which begins or ends in a vertex $f^* \notin H^*$. Then we claim that there exists another operator g' in $C(S(Q))$ such that $\mathrm{supp}(g')$ also contains a path which ends in f^*. If $\mathrm{supp}(g)$ contains a cocycle σ, then consider the operator gh where $h = \prod_{e \in \mathrm{supp}\,\sigma} X_e$. This is also is in $C(S(Q))$ as $h \in C(S(Q))$; further $\mathrm{supp}(gh)$ does not contain σ but it contains the path which ends in f^*. By repeating this process we can assume without loss of of generality that g is a union of paths. By hypothesis there exists some path in $\text{supp}(g)$ that either began or ended in some vertex f^*. This implies that the number of edges in $\text{supp}(g)$ that are incident on f^* must be odd. Consider now the face operator that corresponds to f, namely B_f. It follows that B_f cannot commute with g because it overlaps with g over number of qubits. Thus g is not in $C(S(Q))$, contradicting our assumption. This proves the second part of the lemma.

Before we can prove our result on the local equivalence from stabilizer states derived from planar graphs with punctured discs, we also need the following lemma. This is similar to Lemma 6, the important difference to note is that in this case we are dealing with planar graphs with punctured discs unlike in Lemma 6, wherein we are dealing with graphs on closed surfaces.

Lemma 8. *Let Q be an $[[n, k, d \geq 3]]$ planar surface code derived from a graph Γ without cycles or cocycles of length ≤ 2; we assume that the punctured discs of Γ are given by $H \subsetneq F(\Gamma)$ as in equation (6). Let S be an abelian subgroup of $C(S(Q))$ that contains $S(Q)$. Then the operator $B_f \in S$ is a minimal element of S and $A^S_{\text{supp}(B_f)} = 1$. Further, for any $A_v \in S$ there exist minimal elements $\{h^v_1, \ldots, h^v_{k_v}\} \subset S$ such that:*

i) Each h^v_i is an X-only operator and $\text{supp}(A_v) = \cup^{k_v}_{i=1}\text{supp}(h^v_i)$.
ii) $A^S_{\text{supp}(g^v_i)} = 1$.

Furthermore, for any $j \in \{1, 2, \ldots, n\}$, there exists some A_v and B_f, where $f \in F(\Gamma) \setminus H$, such that $j \in \text{supp}(A_v) \cap \text{supp}(B_f)$, where $n = |E(\Gamma)|$, the total number of qubits.

Proof. We show that every face operator B_f is a minimal element in $C(S(Q))$ and therefore minimal in S also. By Lemma 3, every face $f \in F(\Gamma)$ is bounded by a cycle. Therefore, $\text{supp}(B_f) = \omega$ is a cycle. If B_f is not minimal in $C(S(Q))$, then there exists an element $g \in C(S(Q))$ such that $\emptyset \neq \text{supp}(g) \subsetneq \text{supp}(B_f)$. As Q is a CSS code, this implies $g = g_x g_z$ where g_x (and g_z) are X-only (Z-only) operators respectively. Further, at least one of g_x and g_z is nontrivial with support a proper subset of ω. As g_z is a Z-only operator in $C(S(Q))$, by Lemma 7, $\text{supp}(g_z)$ is a cycle. But a proper subset of a cycle cannot be a cycle. Therefore $g_z = I$, further $g_x \neq I$ if g is not minimal. By Lemma 7, the support of $g_x \in C(S(Q))$ is a union of cycles in Γ^* and/or paths in Γ^*. If $\text{supp}(g_x)$ contains a cocycle then this implies that there exists 2-cocycles in Γ contrary to hypothesis. Therefore $\text{supp}(g_x)$ must be a union of paths in Γ^*. But all the edges of $\text{supp}(g_x)$ are incident on the same vertex in Γ^*. Such paths must necessarily be of length ≤ 2. Since all paths in $\text{supp}(g_x)$ must begin and end in H^* by Lemma 7, it implies that Q has a distance < 3 contrary to assumption. Therefore, B_f is a minimal element.

We now show that $A^S_{\text{supp}(B_f)} = 1$. Suppose that there exists such an element $g \neq B_f$ with $\text{supp}(g) = \text{supp}(B_f)$. Then due to minimality of B_f, both g and B_f must differ on every edge in $\text{supp}(B_f)$. But then g would induce a 2-cocycle in Γ contrary to our assumptions.

We next show that A_v can be written as a linear combination of X-only operators such that $A_v = h_1^v \cdots h_{l_v}^v$ such that $\cup_i \mathrm{supp}(h_i) = \mathrm{supp}(A_v)$. If A_v is minimal then we can simply let $h_1^v = A_v$ and we are done. If not, there exists a minimal element h_1^v within the support of $\mathrm{supp}(A_v)$. The key observation is that h_1^v can be chosen to be an X-only operator because of the fact Q is CSS and because Γ has no cycles of length ≤ 2. The operator $h_1^v A_v$ is such that $\mathrm{supp}(h_1^v) \cup \mathrm{supp}(h_1^v A_v) = \mathrm{supp}(A_v)$. We can then inductively apply the argument to $h_1^v A_v$, $h_2^v h_1^v A_v$, $\ldots$, $h_1^v h_2^v \cdots h_{k_v}^v A_v$.

It is clear that in the absence of loops, every edge is in the support of some vertex operator A_v. Also because there are no coloops, every edge is in the support at exactly two faces. Since no two faces in H share an edge, every edge is in the support of some face $f \in F(\Gamma) \setminus H$. This proves the last part of the lemma.

Theorem 2. *Let Q be an $[[n, k, d \geq 3]]$ planar surface code derived from a 2-connected planar graph Γ and $H \subsetneq F(\Gamma)$. We assume that H satisfies equation (6) and that Γ has no cycles or cocycles of length ≤ 2. Then for any planar surface code (stabilizer) state $|\psi\rangle$ in Q we have $\mathrm{LU}(\psi) = \mathrm{LC}(\psi)$.*

Proof. The proof is very similar to the proof of Theorem 1 so we will only sketch it. First, we note that the stabilizer of $|\psi\rangle$ contains all the vertex operators of Γ, each of which can be expressed a linear combination of X-only minimal elements in $S(\psi)$. Secondly, we note that every face operator $B_f \in S(\psi)$ is minimal. Further, by Lemma 8 the group generated by these minimal elements satisfies Lemma 2. Hence we can conclude that the planar surface code states satisfy $\mathrm{LU}(\psi) = \mathrm{LC}(\psi)$. (The absence of 2-cycles and 2-cocycles in Γ ensures that ψ is free of Bell pairs.)

3.3 Application: On the Existence of Non-clifford Transversal Gates

One of the important techniques to compute fault tolerantly is the use of encoded transversal gates. Although the symmetries of the stabilizer can be used to find transversal gates, such gates are restricted to the Clifford group. Most codes do not permit transversal non-Clifford gates and it remains a challenging problem to find non-Clifford transversal gates. Perhaps the most prominent quantum codes with transversal non-Clifford gates are the quantum Reed-Muller codes [17]. It is now known that a universal set of transversal gates does not exist for stabilizer codes [18]. In fact, they do not exist even for general quantum codes [3]. The restricted problem of finding transversal non-Clifford gates has also been met with stiff resistance, in that Rains [11] has shown that a large and important class of quantum codes, namely the GF(4)-linear quantum codes, do not possess such gates. It remains therefore to find transversal non-Clifford gates for the purely additive quantum codes. However, as we shall show, even under some reasonable restrictions, an important class of additive codes, namely the surface codes cannot possess transversal non-Clifford gates.

Lemma 9. *[18, Lemma 1] Let Q be a stabilizer code. If $U = U_1 \otimes \cdots \otimes U_n$ is a logical gate for Q, then $U_\omega \rho_\omega(Q) = \rho_\omega(Q) U_\omega$ for all $\omega \subseteq \{1, 2, \ldots, n\}$, where $U_\omega = \otimes_{i \in \omega} U_i$,*

$$\rho_\omega = \frac{1}{B_\omega(Q)} \sum_{\substack{s \in S(Q) \\ \operatorname{supp}(s) \subseteq \omega}} \operatorname{Tr}_{\bar\omega}(s); \tag{10}$$

$$B_\omega(Q) = |\{s \in S(Q) : \operatorname{supp}(s) \subseteq \omega\}| \tag{11}$$

for all ω. More generally, if Q' is another stabilizer code such that U is a local equivalence from Q to Q', then $U_\omega \rho_\omega(Q) = \rho_\omega(Q') U_\omega$.

Theorem 3. *Let Q be a surface code such that the associated graph does not have any cycles or cocycles of length ≤ 2. Then Q does not have any transversal encoded non-Clifford gate, $U = U_1 \otimes U_2 \otimes \cdots \otimes U_n$.*

Proof. We assume that Q is derived from a graph Γ. Let g be a site operator or face operator. By Lemma 6, we know that $\operatorname{supp}(g) = \cup_{i \subset k_g} \operatorname{supp}(m_i^g)$, for some minimal elements $\{m_1^g, \ldots, m_{k_g}^g\} \subset S(Q)$. Let $\omega = \operatorname{supp}(m_i^g)$, then we have that $A_\omega^S(Q) = 1$. Now computing ρ_ω as given by (11) we have

$$\rho_\omega = \frac{1}{2} \sum_{\substack{s \in S(Q) \\ \operatorname{supp}(s) \subseteq \omega}} s = \frac{1}{2} \left(\operatorname{Tr}_{\bar\omega}(I) + \operatorname{Tr}_{\bar\omega}(m_i^g) \right).$$

By Lemma 9, $\rho_\omega = U_\omega \rho_\omega U_{\omega^\dagger}$ i.e.,

$$\frac{1}{2} \left(\operatorname{Tr}_{\bar\omega}(I) + \operatorname{Tr}_{\bar\omega}(m_i^g) \right) = \frac{1}{2} \left(\operatorname{Tr}_{\bar\omega}(I) + U_\omega \operatorname{Tr}_{\bar\omega}(m_i^g) U_\omega^\dagger \right)$$

and we obtain $\operatorname{Tr}_{\bar\omega}(m_i^g) = U_\omega \operatorname{Tr}_{\bar\omega}(m_i^g) U_\omega^\dagger$. Since m_i^g has no support outside ω we can conclude that $m_i^g = U m_i^g U^\dagger$.

We have shown so far that any transversal encoded gate of Q maps every minimal operator whose support lies in the support of a site or face operator to itself under conjugation. We now show that this restriction implies that U must be such that every U_i is a Clifford unitary. If U is a non-Clifford encoded gate, then there exists a $j \in \{1, 2, \ldots, n\}$ such that $U_j \notin \mathcal{K}_1$.

By Lemma 6, there exist a site operator A_v and a face operator B_f such that $j \in \operatorname{supp}(A_v) \cap \operatorname{supp}(B_f)$. This implies the existence of an X-only minimal operator g and a Z-only minimal operator h such that $UgU^\dagger = g$ and $UhU^\dagger = h$. Hence,

$$U_j g_j U_j^\dagger \propto g_j \text{ and } U_j h_j U_j^\dagger \propto h_j, \text{ up to a scalar.}$$

Since g is an X-only operator $g_j = X$ and h a Z-only operator $h_j = Z$. This implies that

$$U_j X U_j^\dagger \propto X \text{ and } U_j Z U_j^\dagger \propto Z, \text{ up to a scalar.}$$

Thus $U_j \in \mathcal{K}_1$ contrary to the assumption that it is not in $\mathcal{K}_1$. Thus there exists no transversal encoded non-Clifford gate for the surface codes under the assumptions stated.

It appears that the preceding theorem can be extended to planar surface codes encoding more than one state. In other words, under suitable constraints, we can show that codes derived from planar surfaces with boundaries do not possess transversal non-Clifford encoded gates.

4   Matroids and Surface Code States

Matroids are useful mathematical structures that find applications in many areas such as graph theory, optimization, error-correcting codes, cryptography. It surprising therefore, that matroids have not found such a broad application in quantum information theory. Part of our motivation underlying this section is to explore the use of matroids for quantum information. It turns out that CSS surface code states are naturally associated with matroids. Using the results of Section 3 we associate a matroid with every CSS surface code state. We call matroids arising from CSS surface code states as "surface code matroids". Analogous to the graph minor operations mentioned in Section 2, through which one obtains new graphs from existing graphs, one can obtain new matroids from existing matroids through matroid minor operations which are also called deletion and contraction. We show that the matroids that are obtained by these minor operations are also surface code matroids in that they can be associated with surface code states. The surface code matroids are in this sense minor closed and can be characterized in terms of a list of excluded minors. We refer the reader to [10] for an introduction to matroids.

4.1   Surface Code Matroids

Having defined surface code states, we now associate a matroid to a surface code state $|\psi_\Gamma\rangle$ in a canonical fashion. Recall that a CSS surface code state is stabilized by

$$S(\psi_\Gamma) = \left\langle A_v, B_f, \overline{X}_1, \ldots, \overline{X}_l, \overline{Z}_{l+1}, \ldots, \overline{Z}_k \, \middle| \, \begin{matrix} v \in V(\Gamma) \\ f \in F(\Gamma) \end{matrix} \right\rangle,$$

where we assume that the encoded X operators $\overline{X}_i$ are X-only operators and the encoded Z operators $\overline{Z}_i$ are Z-only operators. In matrix form $S = \begin{bmatrix} S_X & 0 \\ 0 & S_Z \end{bmatrix}$.
Since a CSS stabilizer state is completely determined by S_X or S_Z, we can define the CSS surface code states in terms of $\{A_v \mid v \in V(\Gamma)\} \cup \{\overline{X}_1, \ldots, \overline{X}_l\}$ or $\{B_f \mid f \in F(\Gamma)\} \cup \{\overline{Z}_{l+1}, \ldots, \overline{Z}_k\}$. Let the vertex-edge incidence matrix of Γ be $\mathbb{I}_{V(\Gamma)}$. Denote by $\mathfrak{C}(\Gamma)$ the cycles of Γ. If $B \subseteq \mathfrak{C}(\Gamma)$, the we denote its edge incidence matrix by $\mathbb{I}_B$. The supports of the encoded X and encoded Z operators are cycles in Γ^* and Γ respectively. Denote by $C(\Gamma^*)$, the cycles in Γ^* that correspond to $\{\overline{X}_1, \ldots, \overline{X}_l\}$. We can write S_X in terms of these incidence matrices as

$$S_X = \begin{bmatrix} \mathbb{I}_{C(\Gamma^*)} \\ \mathbb{I}_{V(\Gamma)} \end{bmatrix}, \tag{12}$$

where $C(\Gamma^*) \subseteq \mathfrak{C}(\Gamma^*)$. The surface code matroid of $|\psi_\Gamma\rangle$ is defined as the vector matroid of S_X and we shall denote it as $\mathcal{M}(\psi_\Gamma)$. In other words, $\mathcal{M}(\psi_\Gamma)$ is determined by all the trivial cycles of Γ^* and a subset of the nontrivial cycles of Γ^*. Note that $\mathbb{I}_{V(\Gamma)}$ contains some dependent rows. However, the matroid associated does not change when we add dependent rows to the matrix representing the matroid.

The vector matroid associated with S_Z is the dual matroid of the vector matroid of S_X. Since S_Z is determined by $\{B_f \mid\in F(\Gamma)\} \cup \{\overline{Z}_{l+1}, \ldots, \overline{Z}_k\}$, by duality we see that

$$\mathcal{M}(\psi_\Gamma)^* = \begin{bmatrix} \mathbb{I}_{C'(\Gamma)} \\ \mathbb{I}_{V(\Gamma^*)} \end{bmatrix}, \tag{13}$$

where $C'(\Gamma)$ is a subset of $\mathfrak{C}(\Gamma)$.

Lemma 10. *Let $\mathcal{M}(\psi_\Gamma)$ be a surface code matroid with ground set $E(\Gamma)$. Then any minor of $\mathcal{M}(\psi_\Gamma)$ is also a surface code matroid. Furthermore, $\mathcal{M}(\psi_\Gamma) \setminus e = \mathcal{M}(\psi_{\Gamma \setminus e})$ and $\mathcal{M}(\psi_\Gamma)/e = \mathcal{M}(\psi_{\Gamma/e})$, where $|\psi_{\Gamma \setminus e}\rangle$ and $\psi_{\Gamma/e}$ are some surface code states of $\Gamma \setminus e$ and Γ/e respectively.*

Proof. Since M is a surface code matroid, there exists a graph Γ embedded on some surface Σ, such that the surface code matroid $\mathcal{M}(\psi_\Gamma)$ can be represented as

$$\mathcal{M}(\psi_\Gamma) = \begin{bmatrix} \mathbb{I}_{C(\Gamma^*)} \\ \mathbb{I}_{V(\Gamma)} \end{bmatrix},$$

where $C(\Gamma^*)$ is some subset of the nontrivial (homological) cycles of $\mathfrak{C}(\Gamma^*)$. We shall show that both the matroid deletion and contraction operations on $\mathcal{M}(\psi_\Gamma)$ result in surface code matroids. First let us consider the deletion operation. This corresponds to the deletion of a column of $\mathcal{M}(\psi_\Gamma)$. We show that this is equivalent to the deletion of the edge associated with that column. Without loss of generality assume that the first column is deleted. Assume that this column is associated with the edge e in Γ. Deleting the column corresponding to e, the matrix $I_{V(\Gamma)}$ gives the incidence matrix of $\mathbb{I}_{V(\Gamma \setminus e)}$. Deleting an edge in Γ corresponds to contracting an edge in Γ^*, see Lemma 4. So any cycle $c \in C(\Gamma^*)$ that contains e continues to be a cycle in Γ^*/e unless c is a coloop. If c is a coloop, then contracting e removes the row corresponding to c in $I_{C(\Gamma^*)}$ The cycles in $C(\Gamma^*)$ are also in Γ^*/e which is obtained by contracting the edge e in Γ^*. Denote this subset of cycles in Γ^*/e by $C(\Gamma^*/e)$. In either case deleting the column in $I_{C(\Gamma^*)}$ gives a matrix which is the incidence matrix of cycles in $C(\Gamma^*/e)$. Since $\Gamma^*/e = (\Gamma \setminus e)^*$, $\mathcal{M}(\psi_\Gamma) \setminus e$ can be identified with

$$\mathcal{M}(\psi_\Gamma) \setminus e = \begin{bmatrix} \mathbb{I}_{C(\Gamma^*/e)} \\ \mathbb{I}_{V(\Gamma \setminus e)} \end{bmatrix} = \begin{bmatrix} \mathbb{I}_{C((\Gamma \setminus e)^*)} \\ \mathbb{I}_{V(\Gamma \setminus e)} \end{bmatrix} = \mathcal{M}(\psi_{\Gamma \setminus e}),$$

which shows that the $\mathcal{M}(\psi_\Gamma) \setminus e$ is a surface code matroid.

Next let us consider the matroid contraction of $\mathcal{M}(\psi_\Gamma)$. This corresponds to a projection of the matrix representing $\mathcal{M}(\psi_\Gamma)$ by removing a column as well as a row. If e is not a loop, then it is incident on two vertices u and v and the incidence matrix $\mathbb{I}_{V(\Gamma)}$ contains precisely two rows r_u and r_v that have a '1' in the column corresponding to e. We can replace one of the rows say r_v by their sum $r_u + r_v$. Then this row corresponds to the incidence vector of the vertex obtained by contracting along e. The remaining row r_u corresponds a cycle c in Γ^*. Suppose there is a cycle c' in $C(\Gamma^*)$ such that it contains e, then this cycle can be replaced by another cycle c'' that is obtained by the combination of c and c'. This does not affect $\mathcal{M}(\psi_\Gamma)$, therefore we can assume that all the cycles in $C(\Gamma^*)$ do not contain e. Hence, the cycles in $C(\Gamma^*)$ are also in $\Gamma^* \setminus e$ which is obtained by deleting the edge e in Γ^*. The matroid minor $\mathcal{M}(\psi_\Gamma)/e$ is obtained by removing the row r_u in $\mathcal{M}(\psi_\Gamma)$ and deleting the column corresponding to e, which is now all zero except in the row r_u, because of the elimination operations performed earlier.

Suppose that e is a loop, then the column corresponding to e in $\mathbb{I}_{V(\Gamma)}$ is an all zero column. If all the cycles in $C(\Gamma^*)$ do not contain e, then this column is all zero column and we can simply delete it. Therefore, we have $\mathcal{M}(\psi_\Gamma)/e = \mathcal{M}(\psi_{\Gamma \setminus e}) = \mathcal{M}(\psi_{\Gamma/e})$, where we used the fact that $\Gamma \setminus e = \Gamma/e$, when e is a loop. On the other hand if some cycle $c \in C(\Gamma^*)$ contains e, then we replace every other cycle c' in $C(\Gamma^*)$ that contains e by another cycle c'' such that the row space of $\mathbb{I}_{C(\Gamma^*)}$ does not change. At this point only one cycle in $C(\Gamma^*)$ contains e. Denote these cycles that do not contain e as $C(\Gamma^* \setminus e)$. We obtain $\mathcal{M}(\psi_\Gamma)/e$ by removing the row corresponding to c and deleting the column corresponding to e.

Whether e is loop or not, on contracting the edge e, $\mathbb{I}_{V(\Gamma)}$ gives $\mathbb{I}_{V(\Gamma/e)}$, the vertex-edge incidence matrix of Γ/e, while $\mathbb{I}_{C(\Gamma^*)}$ gives $\mathbb{I}_{C(\Gamma^* \setminus e)}$, cycle-edge incidence matrix of $C(\Gamma^* \setminus e)$. Thus the matroid minor $\mathcal{M}(\psi_\Gamma)/e$ is given by

$$\mathcal{M}(\psi_\Gamma)/e = \begin{bmatrix} \mathbb{I}_{C(\Gamma^* \setminus e)} \\ \mathbb{I}_{V(\Gamma/e)} \end{bmatrix} = \begin{bmatrix} \mathbb{I}_{C((\Gamma/e)^*)} \\ \mathbb{I}_{V(\Gamma/e)} \end{bmatrix} = \mathcal{M}(\psi_{\Gamma/e}).$$

This completes the proof that the minor of a surface code matroid is also a surface code matroid.

A family of matroids $\mathcal{F}$ is said to be minor closed, if every minor of a matroid in $\mathcal{F}$ is also in $\mathcal{F}$. The preceding Lemma, therefore gives us the following result.

Corollary 1. *Surface code matroids are minor closed family of matroids.*

Through Corollary 1 we establish a connection with an important problem in matroid theory. From the structure theory of binary matroids we know that A consequence of the structure theory of binary matroids is that any class of minor closed binary matroids is characterized by a finite set of excluded minors [5]. Thus all CSS surface code states can be characterized by a finite set of CSS states. A secondary consequence of this result and Theorem 1 is that CSS stabilizer states that are counterexamples to the LU-LC conjecture cannot arise from graphic or cographic matroids. The interested reader can find further details in [13].

Acknowledgment

We thank Jim Geelen, and Markus Grassl for useful discussions and the reviewers for their helpful comments. Some of the results of this paper have been presented at the Workshop on Applications of Matroid Theory and Combinatorial Optimization to Information and Coding Theory, Banff International Research Station, Banff, 2009 and an expanded version is available at [13]. This research is supported by CIFAR, IARPA, MITACS and NSERC.

References

1. Bravyi, S., Kitaev, A.Y.: Quantum codes on a lattice with boundary (1998), eprint: quant-ph/9811052
2. Diestel, R.: Graph Theory. Springer, New York (2005)
3. Eastin, B., Knill, E.: Restrictions on transversal encoded quantum gate sets. Phys. Rev. Lett. 102(110502) (2009)
4. Freedman, M.H., Meyer, D.A.: Projective plane and planar quantum codes. Found. Comput. Math. 1(3), 325–332 (2001)
5. Geelen, J.: Binary matroid minors. Applications of Matroid Theory and Combinatorial Optimization to Information and Coding Theory (2009),
   http://robson.birs.ca/~09w5103/geelen_09w5103_talk.pdf
6. Hein, M., Eisert, J., Briegel, H.J.: Multiparty entanglement in graph states. Physical Review A 69(062311) (2004)
7. Ji, Z., Chen, J., Wei, Z., Ying, M.: The LU–LC conjecture is false (2008), arXiv:0709.1266
8. Kitaev, A.Y.: Fault-tolerant quantum computation by anyons. Annals. of Physics 303, 2–30 (2003)
9. Kochen, S., Specker, E.P.: The problem of hidden variables in quantum mechanics. J. Math. Mech. 17(59) (1967)
10. Oxley, J.: What is a matriod? (2004),
    http://www.math.lsu.edu/~oxley/survey4.pdf
11. Rains, E.M.: Quantum codes of minimum distance two. IEEE Trans. Inform. Theory 45(1), 266–271 (1999)
12. Raussendorf, R.: Quantum computation, discreteness and contextuality (2009), arXiv:0907.5449
13. Sarvepalli, P., Raussendorf, R.: On local equivalence, surface code states, and matroids. Physical Review A 82(022304) (2010)
14. Schlingemann, D.: Local equivalence of graph states. In: Krueger, O., Werner, R.F. (eds.) Some Open Problems in Quantum Information Theory (2005), arXiv:quant-ph/0504166
15. van den Nest, M., Dehaene, J., De Moor, B.: An efficient algorithm to recognize local Clifford equivalence of graph states. Physical Review A 70(034302) (2004)
16. van den Nest, M., Dehaene, J., De Moor, B.: Local unitary versus local Clifford equivalence of stabilizer states. Physical Review A 71(062323) (2005)
17. Zeng, B., Chung, H., Cross, A.W., Chuang, I.L.: Local unitary versus local Clifford equivalence of stabilizer and graph states. Physical Review A 75(032325) (2007)
18. Zeng, B., Cross, A., Chuang, I.L.: Transversality and universality for additive quantum codes (2007), arXiv:0706.1382

Testing Non-isometry Is **QMA**-Complete

Bill Rosgen

Centre for Quantum Technologies, National University of Singapore
bill.rosgen@nus.edu.sg

Abstract. Determining the worst-case uncertainty added by a quantum circuit is shown to be computationally intractable. This is the problem of detecting when a quantum channel implemented as a circuit is close to a linear isometry, and it is shown to be complete for the complexity class QMA of verifiable quantum computation. The main idea is to relate the problem of detecting when a channel is close to an isometry to the problem of determining how mixed the output of the channel can be when the input is a pure state.

1 Introduction

A linear isometry $U \colon \mathcal{H} \to \mathcal{K}$ is a linear map that preserves the inner product of any two elements, or equivalently satisfies $U^*U = \mathbb{1}_{\mathcal{H}}$. These transformations are fundamental in quantum computation: they are exactly the maps that may be realized using unitary quantum circuits with access to ancillary qubits in a known pure state—the standard model of quantum computation. It is an important problem to determine when a computation in a non-unitary model, such as measurement based quantum computing or computation in the presence of noise, approximately implements some operation in the unitary circuit model. In this paper it is shown that this problem is QMA-complete when the input computation is modelled as a quantum circuit consisting of the usual unitary gates, plus the ability to discard qubits as well as introduce ancillary qubits. The circuit model is not essential: the hardness result also applies to any model that can efficiently simulate the mixed-state circuit model.

The complexity class QMA is the quantum analogue of NP: the class corresponding to classically verifiable computation. This concept was first considered in [11], first formally defined in [9], and first studied in [18]. QMA is the class of all problems that can be verified with bounded error by a polynomial-time quantum verifier with access to a quantum proof. This proof is given by a quantum state on a polynomial number of qubits and may depend on the input.

The class QMA has complete (promise) problems: problems in QMA that are computationally at least as hard as any other problem in the class. This implies that an efficient algorithm for any of these complete problems can be used to find an efficient algorithm for any problem in QMA. The simplest of these complete problems is the 2-local Hamiltonian problem, which is informally the quantum version of the circuit satisfiability problem for unitary circuits with gates of constant size. A formal description of this problem, as well as a proof

W. van Dam et al. (Eds.): TQC 2010, LNCS 6519, pp. 63–76, 2011.

that the 5-local Hamiltonian problem is QMA-complete can be found in [10]. The improvement of this result to the 2-local case is due to Kempe, Kitaev, and Regev [8]. Several other complete problems for QMA are known, such as local consistency [12] (see also [13,19]), some problems related to the minimum output entropy [2], testing whether unitary circuits are close to the identity [6] (see also [7]), and finding the ground states of some physical systems [17,16]. In the present paper we add a new complete problem to this list: the problem of determining if a quantum circuit implements an operation that is close to an isometry. As discussed in Section 3, this is equivalent to determining if the channel always maps pure states to states that are approximately pure.

The remainder of the paper is organized as follows. Section 2 introduces notation and background. Section 3 introduces the notion of approximate isometries and makes formal the problem of detecting when a channel is an approximate isometry. The QMA-hardness of this problem is proved in Section 4 and proof of the containment in QMA, the most technical portion of the result, appears in Section 5.

2   Preliminaries

In this section the notation and background that is used throughout the paper are presented. Much of the notation used here is standard and this is in no way a complete introduction to quantum information. See [15] for a more detailed treatment of these topics.

All Hilbert spaces considered in this paper are assumed to be finite-dimensional and are denoted by scripted capital letters $\mathcal{H}, \mathcal{K}, \ldots$. The pure states are the unit vectors in these spaces. The set of density matrices or mixed states on $\mathcal{H}$ is given by $\mathbf{D}(\mathcal{H})$, and the set of all quantum channels mapping $\mathbf{D}(\mathcal{H})$ to $\mathbf{D}(\mathcal{K})$ is $\mathbf{T}(\mathcal{H}, \mathcal{K})$. The quantum channels are exactly the completely positive and trace preserving linear maps. The identity channel in $\mathbf{T}(\mathcal{H}, \mathcal{H})$ is denoted $I_{\mathcal{H}}$, while $\mathbb{1}_{\mathcal{H}}$ is the identity on $\mathcal{H}$.

Given a quantum channel $\Phi \in \mathbf{T}(\mathcal{H}, \mathcal{K})$ we make use of two representations. The first of these is the Choi representation [4], which provides a unique representation of a channel $\Phi \in \mathbf{T}(\mathcal{H}, \mathcal{K})$ as a linear operator on $\mathcal{K} \otimes \mathcal{H}$. This representation is given by $\mathrm{C}(\Phi) = (\Phi \otimes I_{\mathcal{H}})(|\phi^+\rangle\langle\phi^+|)$, where $|\phi^+\rangle = \sum_i |ii\rangle/\sqrt{d}$ is a maximally entangled state in $\mathcal{H} \otimes \mathcal{H}$.

The second representation that we use is the representation of a completely positive map Φ by a set of Kraus operators: matrices A_i such that $\Phi(X) = \sum_i A_i X A_i^*$. This representation is also due to Choi [4]. If in addition the map Φ is trace preserving, then the operators A_i satisfy the property $\sum_i A_i^* A_i = \mathbb{1}$. The number of Kraus operators in a minimal Kraus decomposition is given by the rank of the Choi matrix $\mathrm{C}(\Phi)$.

In order to measure how close a state is to being pure we use the operator norm $\|X\|_\infty$, which for a linear operator X is the largest singular value of X. When X is normal, this is simply the largest eigenvalue (in absolute value) of X. Dual to the operator norm is the trace norm, which for a linear operator X

is given by $\|X\|_{\mathrm{tr}} = \mathrm{tr}\,\sqrt{X^*X}$. This is exactly the sum of the singular values of X. When X is a quantum state, this simplifies to the sum of absolute values of the eigenvalues of X, so that $\|\rho\|_{\mathrm{tr}} = 1$ for all density matrices ρ.

One final quantity that we use is the fidelity, which for two density matrices ρ, σ is given by $\mathrm{F}(\rho, \sigma) = \mathrm{tr}\,\sqrt{\sqrt{\rho}\sigma\sqrt{\rho}}$. While it is not obvious from this definition, the fidelity is symmetric in the two arguments. When one of the arguments is a pure state, the fidelity simplifies to $\mathrm{F}(\rho, |\psi\rangle\langle\psi|) = \sqrt{\langle\psi|\rho|\psi\rangle}$. An important relationship between the trace norm and the fidelity is

$$2 - 2\,\mathrm{F}(\rho, |\psi\rangle\langle\psi|)^2 \leq \|\rho - |\psi\rangle\langle\psi|\|_{\mathrm{tr}},$$

which we use to relate different notions of the purity of a quantum state. This inequality can be found in [15, Chapter 9].

We require one final piece of background. In order for a quantum channel to be given as input to a computational problem we need a representation of the channel. Using either the Choi matrix or Kraus operators produces a representation that, in the case of channels implementing efficient quantum algorithms, is exponentially larger than the size of a circuit representation. These channels have circuit representations that are logarithmic in the Hilbert space dimension. For this reason, we use a circuit representation of quantum channels. Such a representation is provided by the mixed-state circuit model of Aharonov et al. [1], which is simply the usual model of unitary quantum circuits with two additional gates. These gates are the gate that introduces ancillary qubits in the $|0\rangle$ state and the gate the traces out (i.e. discards) a qubit. This circuit model can be used to represent any quantum channel, which makes it ideal for the problem that we consider.

3   Isometries and Rank Non-increasing Channels

One important property of the linear isometries is that they do not increase rank. This is essential to the QMA protocol in Section 5, which is able to detect exactly those channels that are rank-increasing. More formally, a channel Φ is *rank non-increasing* if for all states ρ the output of Φ satisfies $\mathrm{rank}(\rho) \geq \mathrm{rank}(\Phi(\rho))$. Unfortunately, this property does not characterize the isometries. Consider the channel $\Phi(\rho) = |0\rangle\langle0|$ that discards the input state and returns a fixed pure state. This channel is not an isometry but it is also rank non-increasing.

This property can be used to characterize the isometries if we make a small adjustment. The channels that are rank non-increasing when adjoined to an auxiliary space of arbitrary dimension are exactly the isometries. We call a channel $\Phi \in \mathbf{T}(\mathcal{H}, \mathcal{K})$ *completely* rank non-increasing if for any auxiliary space $\mathcal{F}$ the channel $\Phi \otimes I_{\mathcal{F}}$ is rank non-increasing, i.e. if $\mathrm{rank}\,[(\Phi \otimes I_{\mathcal{F}})(\rho)] \leq \mathrm{rank}(\rho)$ for all ρ. The channel $\Phi(\rho) = |0\rangle\langle0|$ is not completely rank non-increasing: consider applying it to half of a maximally entangled state $(\Phi \otimes I_{\mathcal{H}})(|\phi^+\rangle\langle\phi^+|) = |0\rangle\langle0| \otimes \mathbb{1}_{\mathcal{H}}/\dim\mathcal{H}$. As in the case of complete positivity, we need only to verify this property on an auxiliary space of the same dimension as the input space. It is also easy to see that this property characterizes the linear isometries.

Proposition 1. *The following are equivalent for a channel $\Phi \in \mathbf{T}(\mathcal{H}, \mathcal{K})$:*

1. *$\Phi(\rho) = U\rho U^*$ for some linear isometry U from $\mathcal{H}$ to $\mathcal{K}$,*
2. *Φ is completely rank non-increasing,*
3. *$\Phi \otimes I_{\mathcal{H}}$ is rank non-increasing.*

Proof. The first two implications are immediate. To prove that $(3) \Rightarrow (1)$, let $\Phi \otimes I_{\mathcal{H}}$ be rank non-increasing. This implies that $\text{rank}(C(\Phi)) = 1$. Recalling that the number of Kraus operators in a minimal decomposition is $\text{rank}(C(\Phi))$, it follows that Φ can be expressed as $\Phi(\rho) = A\rho A^*$. The condition that Φ is trace preserving implies that the operator A satisfies $A^*A = \mathbb{1}_{\mathcal{H}}$. $\square$

This characterization guides the remainder of the paper. Detecting when the channel $\Phi \otimes I_{\mathcal{H}}$ increases rank provides an operational method to determine when a channel is an isometry.

3.1   Approximately Pure States

In order to show that non-isometry detection is QMA-complete we need to consider an approximate version of the problem. This is because a protocol for a QMA language is permitted to fail with small probability. The definition of approximate isometries used here is closely related to the notion of approximately pure states. Several equivalent notions of the purity of a density matrix are considered in this section.

Perhaps the most well-known notion of how close a mixed state ρ is to being pure is the *purity* of ρ, given by $\text{tr}(\rho^2)$. A similar measure is given by $\|\rho\|_\infty$, the largest eigenvalue of ρ. It is not hard to see that these quantities are related. If $\rho = \sum_i \lambda_i |\lambda_i\rangle\langle\lambda_i|$ is the spectral decomposition of ρ, with the eigenvalues λ_i in decreasing order, then $\text{tr}\,\rho^2 = \sum_i \lambda_i^2 \geq \lambda_1^2 = \|\rho\|_\infty^2$. In the other direction, since the purity is convex, it is maximized for $1/\lambda_1$ eigenvalues each of value λ_1, i.e. $\text{tr}\,\rho^2 = \sum_i \lambda_i^2 \leq \lambda_1^2/\lambda_1 = \|\rho\|_\infty$. Taken together, these two inequalities show that

$$\|\rho\|_\infty^2 \leq \text{tr}(\rho^2) \leq \|\rho\|_\infty . \tag{1}$$

These quantities are also related to the more familiar trace distance on quantum states.

Proposition 2. *Let $\rho \in \mathbf{D}(\mathcal{H})$ and let $\varepsilon > 0$. There exists a pure state $|\psi\rangle \in \mathcal{H}$ such that $\|\rho - |\psi\rangle\langle\psi|\|_{\text{tr}} \leq \varepsilon$ if and only if $\|\rho\|_\infty \geq 1 - \varepsilon/2$.*

Proof. Let ρ have spectral decomposition given by $\rho = \sum_i \lambda_i |\lambda_i\rangle\langle\lambda_i|$, with $\lambda_1 \geq \lambda_2 \geq \ldots \geq \lambda_d$. If $\|\rho\|_\infty = \lambda_1 \geq 1 - \varepsilon/2$, then

$$\|\rho - |\lambda_1\rangle\langle\lambda_1|\|_{\text{tr}} = (1 - \lambda_1) + \sum_{i=2}^{d} \lambda_i = 2(1 - \lambda_1) \leq 2(\varepsilon/2) = \varepsilon.$$

On the other hand, if $|\psi\rangle \in \mathcal{H}$ is a state such that $\|\rho - |\psi\rangle\langle\psi|\|_{\text{tr}} \leq \varepsilon$, then

$$\varepsilon \geq \|\rho - |\psi\rangle\langle\psi|\|_{\text{tr}} \geq 2 - 2\,\text{F}(\rho, |\psi\rangle\langle\psi|)^2$$
$$= 2 - 2\langle\psi|\rho|\psi\rangle = 2 - 2\sum_i \lambda_i |\langle\psi|\lambda_i\rangle|^2 . \tag{2}$$

The final quantity is a convex combination of the λ_i, with weights determined by the state $|\psi\rangle$. This is maximized when $|\psi\rangle = |\lambda_1\rangle$, since λ_1 is the largest eigenvalue of ρ. Combining this with Equation (2) we have $\varepsilon \geq 2 - 2\lambda_1 = 2 - 2\|\rho\|_\infty$, which implies that $\|\rho\|_\infty \geq 1 - \varepsilon/2$. $\square$

Given these notions of purity, we call a state ε-*pure* if $\|\rho\|_\infty \geq 1 - \varepsilon$. By the above results the purity of such a state satisfies $\operatorname{tr}(\rho^2) \geq (1 - \varepsilon)^2 \geq 1 - 2\varepsilon$ and there is a pure state $|\psi\rangle$ such that $\|\rho - |\psi\rangle\langle\psi|\|_{\mathrm{tr}} \leq 2\varepsilon$. For the results of this paper, any of these three measures suffices, as they are equivalent up to polynomial factors in ε.

3.2   Approximate Isometries

The focus of this paper is to show that detecting when a channel is far from an isometry is computationally difficult. To do this we need to define the class of channels that are the approximate isometries. Isometries always map pure states to pure states, even in the presence of a reference system. Proposition 1 shows that this condition characterizes the isometries. Weakening this requirement, we call a channel an ε-*isometry* if it maps pure states (over the input space and a reference system) to states that are ε-pure, for some $\varepsilon > 0$.

More formally a channel $\Phi \in \mathbf{T}(\mathcal{H}, \mathcal{K})$ is an ε-isometry if for any pure state $|\psi\rangle \in \mathcal{H} \otimes \mathcal{H}$ the output of $\Phi \otimes I_\mathcal{H}$ satisfies $\|(\Phi \otimes I_\mathcal{H})(|\psi\rangle\langle\psi|)\|_\infty \geq 1 - \varepsilon$, i.e. when applied to part of any pure state the output state is close to pure. This implies that $\Phi \otimes I_\mathcal{H}$ does not reduce the operator norm of any input by more than a factor of $1 - \varepsilon$. We use this to define the computational problem that is the main focus of the paper.

Problem (Non-isometry). For $0 \leq \varepsilon < 1/2$ and $\Phi \in \mathbf{T}(\mathcal{H}, \mathcal{K})$, given as a mixed-state quantum circuit, the promise problem is to decide between:

Yes: There exists $|\psi\rangle \in \mathcal{H}$ such that $\|(\Phi \otimes I_\mathcal{H})(|\psi\rangle\langle\psi|)\|_\infty \leq \varepsilon$,
No: For all $|\psi\rangle \in \mathcal{H}$, $\|(\Phi \otimes I_\mathcal{H})(|\psi\rangle\langle\psi|)\|_\infty \geq 1 - \varepsilon$.

When ε is significant we refer to this problem as $\textsc{Non-isometry}_\varepsilon$.

Using the equivalence results of Equation (1) and Proposition 2, this problem may be equivalently defined in terms of either the purity or the trace distance to the closest pure state, up to a small increase in ε. The case of the minimum output purity of a channel has been studied in a different context by Zanardi and Lidar [20], though they focus on finding the minimum purity of a channel over a subspace of the inputs. The problem we consider here is equivalent to evaluating the channel purity of $\Phi \otimes I_\mathcal{H}$ over the whole input space.

The difficulty of the $\textsc{Non-isometry}$ problem does not change if the dimension of the ancillary system is permitted to be larger than the size of the input system, so long as the number of qubits needed to represent the ancillary system is polynomial in the number of input qubits.

The notion of approximate isometry that we consider here is *not* equivalent to the channel being completely rank non-increasing on average. This property is modelled by the distance between the Choi matrix of a channel and a pure state. While it is true that the Choi matrix is pure if and only if the channel is an isometry, it is close to pure in the trace distance when the channel is close to an isometry *on average*. In this paper we consider the worst-case, i.e. we consider a channel to be close to an isometry if and only if the output of $\Phi \otimes I_{\mathcal{H}}$ is close to pure for *any* pure state input. A simplification of the protocol presented in Section 5 yields a polynomial-time quantum algorithm for the problem of determining how close the Choi matrix of a channel is to a pure state. This is because $C(\Phi)$ can be generated efficiently, and given two copies the swap test can be used to test the purity of a quantum state as shown in [5].

4   QMA Hardness

In order to prove the hardness of NON-ISOMETRY we modify an arbitrary QMA protocol to obtain a circuit that can output a mixed state exactly when the verifier would have accepted in the original protocol. This yields a circuit that is far from an isometry if and only if there is a witness that causes the verifier in the original protocol to accept. Deciding whether or not there is such a witness is QMA-hard, by the definition of the complexity class. More formally, a language L is in QMA if there is a quantum polynomial-time verifier V such that

1. if $x \in L$, there exists a witness $|\psi\rangle$ such that $\Pr[V$ accepts $|\psi\rangle] \geq 1 - \varepsilon$,
2. if $x \notin L$, then for any state $|\psi\rangle$, $\Pr[V$ accepts $|\psi\rangle] \leq \varepsilon$,

The exact value of the error parameter ε is not significant: any $\varepsilon < 1/2$ that is at least an inverse polynomial in the input size suffices [10,14].

Let L be an arbitrary language in QMA, and let x be an arbitrary input string. The goal is to embed the QMA-hard problem of deciding if $x \in L$ into the problem of testing whether a mixed-state quantum circuit is close to an isometry. Let V be the isometry representing the algorithm of the verifier in a QMA protocol for L on input x. We may "hard-code" the input string x into V because the circuit needs only to be efficiently generated from x. The algorithm implemented by the verifier is shown in Figure 1. The verifier first receives a witness state $|\psi\rangle$, applies the isometry V, and then makes a measurement on one of the qubits, the result of which determines whether or not the verifier accepts. Any qubits not measured are traced out.

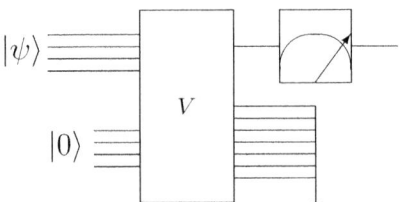

Fig. 1. Verifier's circuit in a QMA protocol. The verifier accepts the witness state $|\psi\rangle$ if and only if the measurement in the computational basis results in the $|1\rangle$ state.

For concreteness, let V act on the input spaces $\mathcal{W}$ and $\mathcal{A}$, which hold the witness state and the $|0\rangle$ state of the ancilla respectively. Let $\mathcal{M}$ be the space corresponding to the measured output qubit in the protocol and let $\mathcal{G}$ represent the 'garbage' qubits that are traced out at the end of the protocol. The probability that verifier accepts the witness state $|\psi\rangle \in \mathcal{W}$ is

$$\Pr[V \text{ accepts } |\psi\rangle] = \langle 1| \operatorname{tr}_{\mathcal{G}} \left[V(|\psi\rangle\langle\psi| \otimes |0\rangle\langle 0|) V^* \right] |1\rangle. \tag{3}$$

Deciding if there is some $|\psi\rangle$ such that this expectation is close to one is complete for QMA.

From Figure 1 it is simple to construct a circuit that produces highly mixed output exactly when there exists such a $|\psi\rangle$. The idea is add a controlled application of the completely depolarizing channel Ω on the space $\mathcal{G}$, instead of tracing it out. The resulting circuit is shown in Figure 2. In the case that the verifier accepts with negligible probability for every input state $|\psi\rangle$, then both the measurement and the controlled depolarizing channel have little effect, leaving the state of the system close to a pure state. If, on the other hand, there is a state on which the verifier accepts with high probability, then on this input the circuit in Figure 2 produces a highly mixed state. Formalizing this notion proves that NON-ISOMETRY is QMA-hard.

Theorem 3. *Let $\varepsilon > 0$ be a constant, and let p be the maximum acceptance probability of the protocol V. Let $\Phi \in \mathbf{T}(\mathcal{W}, \mathcal{M} \otimes \mathcal{G})$ be the circuit in Figure 2. Then if $\dim \mathcal{R} = \dim \mathcal{W}$*

$$p \le \varepsilon \implies \min_{|\psi\rangle} \|(\Phi \otimes I_{\mathcal{R}})(|\psi\rangle\langle\psi|)\|_\infty \ge 1 - \varepsilon,$$

$$p \ge 1 - \varepsilon \implies \min_{|\psi\rangle} \|(\Phi \otimes I_{\mathcal{R}})(|\psi\rangle\langle\psi|)\|_\infty \le \varepsilon.$$

Proof. Notice that we may assume that the output dimension of Φ is $\dim \mathcal{M} \otimes \mathcal{G} = 2d > 2/\varepsilon$ by padding the circuit for V with $\log 1/\varepsilon$ unused ancillary qubits, if necessary.

As a first step, we evaluate the output state of the channel $\Phi \otimes I_{\mathcal{R}}$. Applied to a pure state $|\psi\rangle \in \mathcal{W} \otimes \mathcal{R}$ this channel first adds the ancillary $|0\rangle$ qubits in the

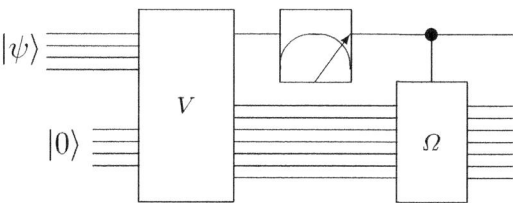

Fig. 2. Constructed instance of NON-ISOMETRY. The output state is mixed by the completely depolarizing channel Ω only if the state $|\psi\rangle$ is a valid witness to the original QMA protocol.

space $\mathcal{A}$ and then applies the isometry V from the QMA protocol. This results in the pure state $|\phi\rangle = (V \otimes \mathbb{1}_{\mathcal{R}})(|\psi\rangle \otimes |0\rangle)$. We may decompose this state in terms of the qubit in the space $\mathcal{M}$, obtaining for some $0 \leq p \leq 1$

$$|\phi\rangle = \sqrt{1-p}|0\rangle|\phi_0\rangle + \sqrt{p}|1\rangle|\phi_1\rangle.$$

The value of p is exactly the probability that the measurement result is $|1\rangle$, i.e. the probability that the verifier will accept the input state $\mathrm{tr}_{\mathcal{R}} |\psi\rangle\langle\psi|$ in the original protocol. Using this, the state after the measurement and the controlled depolarizing channel on $\mathcal{G}$ is

$$(1-p)|0\rangle\langle 0| \otimes |\phi_0\rangle\langle\phi_0| + (p/d)|1\rangle\langle 1| \otimes \mathbb{1}_{\mathcal{G}} \otimes \rho, \qquad (4)$$

where ρ is the residual state on $\mathcal{R}$ after this channel has been applied ($\rho = \mathrm{tr}_{\mathcal{G}} |\phi_1\rangle\langle\phi_1|$, but this will not be important). Evaluating the largest eigenvalue of this state we find that

$$\|(\Phi \otimes I_{\mathcal{R}})(|\phi\rangle\langle\phi|)\|_{\infty} = \max\{1-p, \frac{p}{d}\|\rho\|_{\infty}\}. \qquad (5)$$

We analyze the maximum in Equation (5) in two cases. The first of these cases is when there is no input the verifier accepts with probability larger than ε. In this case the output of the channel $\Phi \otimes I_{\mathcal{R}}$ is given by Equation (4) where $p \leq \varepsilon$. Here Equation (5) shows that the output satisfies $\min_{|\mu\rangle} \|(\Phi \otimes I_{\mathcal{R}})(|\mu\rangle\langle\mu|)\|_{\infty} \geq 1 - p \geq 1 - \varepsilon$.

The second case is when there exists a state $|\psi\rangle$ that verifier to accepts with probability at least $1 - \varepsilon$. In this case we take the input state to $\Phi \otimes I_{\mathcal{R}}$ to be $|\gamma\rangle = |\psi\rangle \otimes |0\rangle$, i.e. we set the reference system to be any pure state that is not entangled with the rest of the input. The output is given by Equation (4) with $p \geq 1 - \varepsilon$ and $\rho = |0\rangle\langle 0|$. Equation (5) yields

$$\min_{|\mu\rangle} \|(\Phi \otimes I_{\mathcal{R}})(|\mu\rangle\langle\mu|)\|_{\infty} \leq \|(\Phi \otimes I_{\mathcal{R}})(|\gamma\rangle\langle\gamma|)\|_{\infty}$$

$$= \max\left\{1-p, \frac{p}{d}\|\rho\|_{\infty}\right\} \leq \max\left\{\varepsilon, \frac{1}{d}\right\} = \varepsilon,$$

as we have taken $1/d < \varepsilon$ (by adding $O(\log 1/\varepsilon)$ unused ancillary qubits if necessary). □

This theorem shows that determining how far the output $\Phi \otimes I_{\mathcal{R}}$ is from a pure state is as computationally difficult as determining whether or not the verifier can be made to accept with high probability in a QMA protocol. Since the construction of the circuit shown in Figure 2 can be performed efficiently, this implies the hardness of this problem.

Corollary 4. *For any constant $0 \leq \varepsilon < 1/2$, NON-ISOMETRY is QMA-hard.*

Using the equivalences between notions of purity in of Section 3.1, this also implies that evaluating the purity of a quantum channel, as defined by Zanardi and Lidar [20], is QMA-hard.

5   QMA Protocol

In order to show that NON-ISOMETRY is QMA-complete, it remains only to construct a QMA protocol for the problem. The key idea behind this protocol is that when two copies of a channel Φ are applied in parallel to the input state $|\psi\rangle \otimes |\psi\rangle$ the output lies in the antisymmetric subspace if and only if $\Phi(|\psi\rangle\langle\psi|)$ is a mixed state. This provides a probabilistic test that can detect when a channel is far from an isometry.

Unfortunately, in a QMA protocol the verifier cannot assume the witness is given by two non-entangled pure states. It suffices, however, for the verifier to require that the input state lies in the symmetric subspace of the input space $(\mathcal{H} \otimes \mathcal{R})^{\otimes 2}$. To show that the channel is not an isometry in QMA, the prover can provide a symmetric state that a parallel application of the channel maps into the antisymmetric space of the output space $(\mathcal{K} \otimes \mathcal{R})^{\otimes 2}$.

The verifier in such a protocol needs a test to determine when a state is symmetric or antisymmetric. Such a test is provided by the swap test, which was introduced in the context of communication complexity in [3], though we make use of it to test purity using an idea from [5]. The swap test can be characterized as the projection onto the symmetric and antisymmetric subspaces of a bipartite space. If W is the swap operation on a space $\mathcal{H} \otimes \mathcal{H}$, then the symmetric measurement outcome of the swap test corresponds to the the projector $(\mathbb{1}_{\mathcal{H}\otimes\mathcal{H}} + W)/2$, and the projector $(\mathbb{1}_{\mathcal{H}\otimes\mathcal{H}} - W)/2$ corresponds to the antisymmetric outcome.

The main idea behind the protocol for NON-ISOMETRY is that the swap test can be used to measure the purity of a state. As observed in [5], when applied two to copies of a state $\rho = \sum_i \lambda_i |\psi_i\rangle\langle\psi_i|$ the swap test returns the antisymmetric outcome with probability

$$\frac{1}{2} \operatorname{tr}((\mathbb{1} - W)(\rho \otimes \rho)) = \frac{1}{2} - \frac{1}{2} \sum_{ij} \lambda_i \lambda_j \operatorname{tr}\left[W(|\psi_i\rangle\langle\psi_i| \otimes |\psi_j\rangle\langle\psi_j|)\right]$$

$$= \frac{1}{2} - \frac{1}{2} \sum_i \lambda_i^2 = \frac{1}{2} - \frac{1}{2} \operatorname{tr}(\rho^2). \tag{6}$$

This implies that the swap test on two copies of a state can be used to test purity and, by extension, when a channel is far from an isometry.

A straightforward protocol for NON-ISOMETRY on a channel Φ is then to receive a witness state $|\psi\rangle \otimes |\psi\rangle$, apply the channel in parallel to obtain $[(\Phi \otimes I)(|\psi\rangle\langle\psi|)]^{\otimes 2}$, and finally apply the swap test. The result is the antisymmetric outcome with high probability only when the output state $(\Phi \otimes I)(|\psi\rangle\langle\psi|)$ is highly mixed. Such a protocol detects the channels that are far from isometries.

Unfortunately, the verifier in a QMA protocol cannot assume that the witness state is of the form $|\psi\rangle \otimes |\psi\rangle$. The verifier *can* check that he has received some state in the symmetric subspace and then use the fact that this subspace is closed under the parallel application of a rank non-increasing channel. The verifier in the following protocol uses the swap test to both check the symmetry of the input state and the antisymmetry of the output state.

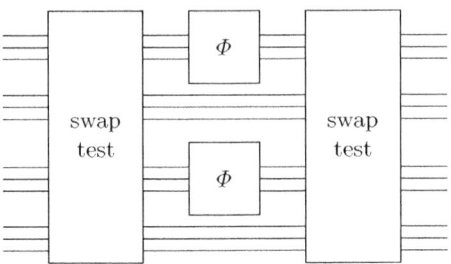

Fig. 3. QMA protocol for NON-ISOMETRY. The verifier accepts only if the first swap test results the symmetric outcome and the second swap test results in an antisymmetric outcome.

Protocol 5 (Non-isometry). On an input channel $\Phi \in \mathbf{T}(\mathcal{H}, \mathcal{K})$:

1. Receive a witness state $\rho \in \mathbf{D}((\mathcal{H} \otimes \mathcal{R})^{\otimes 2})$, where $\mathcal{R}$ is a reference space such that $\dim \mathcal{R} = \dim \mathcal{H}$. Apply the swap test to ρ, rejecting if the outcome is antisymmetric.
2. Use the channel Φ to obtain $\sigma = (\Phi \otimes I_\mathcal{R})^{\otimes 2}(\rho)$.
3. Apply the swap test to σ, accepting if the outcome is antisymmetric and rejecting otherwise.

A diagram of this protocol can be found in Figure 3. The correctness of the protocol is argued in the following theorem.

Theorem 6. *Let $\Phi \in \mathbf{T}(\mathcal{H}, \mathcal{K})$, and let $p(\rho)$ be the probability that the verifier described in Protocol 5 accepts the input state $\rho \in \mathbf{D}((\mathcal{H} \otimes \mathcal{R})^{\otimes 2})$, then*

1. *If $\min_{|\psi\rangle} \|(\Phi \otimes I_\mathcal{R})(|\psi\rangle\langle\psi|)\|_\infty \leq \varepsilon$, then there exists a witness state ρ such that $p(\rho) \geq (1 - \varepsilon)/2$.*
2. *If $\min_{|\psi\rangle} \|(\Phi \otimes I_\mathcal{R})(|\psi\rangle\langle\psi|)\|_\infty \geq 1 - \varepsilon$, then for any witness state ρ, $p(\rho) \leq 9\varepsilon$.*

Proof. For the sake of brevity, let $\hat{\Phi} = \Phi \otimes I_\mathcal{R}$ throughout. To prove the first assertion, let $|\psi\rangle$ be a pure state in $\mathcal{H} \otimes \mathcal{R}$ for which the output of the channel satisfies $\|\hat{\Phi}(|\psi\rangle\langle\psi|)\|_\infty \leq \varepsilon$, and let the witness state be $\rho = |\psi\rangle\langle\psi| \otimes |\psi\rangle\langle\psi|$. This state is invariant under the swap operation and so the swap test in Step 1 passes and does not change the state. Step 2 results in the state $[\hat{\Phi}(|\psi\rangle\langle\psi|)]^{\otimes 2}$. Using Equations (1) and (6), the final swap test returns the antisymmetric outcome with probability

$$\frac{1}{2} - \frac{1}{2} \operatorname{tr}\left[\hat{\Phi}(|\psi\rangle\langle\psi|)^2\right] \geq \frac{1}{2} - \frac{1}{2} \left\|\hat{\Phi}(|\psi\rangle\langle\psi|)\right\|_\infty \geq \frac{1 - \varepsilon}{2},$$

and so the verifier accepts ρ with probability approaching one-half for small ε.

To show the second assertion, we take $\hat{\Phi}$ is an ε-isometry and analyze the probability that the verifier can be made to accept. We may assume that the witness state lies in the symmetric subspace of $(\mathcal{H} \otimes \mathcal{R})^{\otimes 2}$, as the verifier either rejects in Step 1 or projects the witness onto this subspace. To complete the proof, we show that $(\hat{\Phi})^{\otimes 2}$ leaves ρ approximately symmetric.

To do this, we approximate $\hat{\Phi}$ by an operator that preserves the symmetry of input states. Let $\{|i\rangle : 1 \leq i \leq \dim \mathcal{H}\}$ be an orthonormal basis for the spaces $\mathcal{H}, \mathcal{R}$ (this is possible because they have the same dimension). The states $\{|ij\rangle : 1 \leq i, j \leq \dim \mathcal{H}\}$ are an orthonormal basis for $\mathcal{H} \otimes \mathcal{R}$. Since $\hat{\Phi}$ approximately preserves rank, there are states $|\psi_i\rangle \in \mathcal{K}$ such that

$$\| (\Phi \otimes I_\mathcal{R})(|ij\rangle\langle ij|) - |\psi_i\rangle\langle\psi_i| \otimes |j\rangle\langle j| \|_{\text{tr}} \leq \varepsilon \tag{7}$$

for all i and j. We define a linear operator $A : \mathcal{H} \to \mathcal{K}$ by the equation $A|i\rangle = c_i|\psi_i\rangle$, where the $c_i \in \mathbb{C}$ with $|c_i| = 1$. The introduction of the phases c_i is necessary because Equation (7) only defines the states $|\psi_i\rangle$ up to a phase. Note that the operator A is not necessarily unitary as we may not assume that the states $|\psi_i\rangle$ are orthogonal. The next step is to show that, for some choice of the phases c_i, conjugation by A approximates the channel Φ in the trace norm. This is the most technical portion of the proof.

Consider the output of $\hat{\Phi}$ on the entangled state $(|ii\rangle + |jj\rangle)/\sqrt{2}$ in $\mathcal{H} \otimes \mathcal{R}$, given by

$$\rho = \frac{1}{2} \sum_{a,b \in \{i,j\}} \hat{\Phi}(|aa\rangle\langle bb|) = \frac{1}{2} \sum_{a,b \in \{i,j\}} \Phi(|a\rangle\langle b|) \otimes |a\rangle\langle b|. \tag{8}$$

Since $\hat{\Phi}$ maps pure states to states that are nearly pure, we know that the purity of ρ satisfies $\text{tr}(\rho^2) \geq (1-\varepsilon)^2 \geq 1-2\varepsilon$. Evaluating the purity using Equation (8) gives

$$1 - 2\varepsilon \leq \text{tr}(\rho^2) = \frac{1}{4} \left(\text{tr}\,\Phi(|i\rangle\langle i|)^2 + \text{tr}\,\Phi(|j\rangle\langle j|)^2 + 2\,\text{tr}\,\Phi(|i\rangle\langle j|)\Phi(|j\rangle\langle i|) \right)$$

$$\leq \frac{1}{2} + \frac{1}{2}\,\text{tr}\left((\Phi(|i\rangle\langle j|)\Phi(|i\rangle\langle j|)^*) \right). \tag{9}$$

Interpreting the expression $\text{tr}\,XX^*$ as the sum of the squared singular values of X, Equation (9) implies that the operator $\Phi(|i\rangle\langle j|)$ has largest singular value at least $1 - 4\varepsilon$. Since the sum of the singular values of this operator cannot exceed one (as the trace norm does not increase under the application of a channel), this implies that it can be decomposed as

$$\Phi(|i\rangle\langle j|) = (1 - 4\varepsilon)|\phi_i\rangle\langle\phi_j| + 4\varepsilon Y, \tag{10}$$

where $|\phi_i\rangle, |\phi_j\rangle \in \mathcal{K}$ are pure and Y is some linear operator on $\mathcal{K}$ with $\|Y\|_{\text{tr}} = 1$. It remains to show that the vectors $|\phi_i\rangle$ and $|\phi_j\rangle$ are, up to a phase, approximately equal to the vectors $|\psi_i\rangle$ and $|\psi_j\rangle$ defined in Equation (7). To do this, we consider the action of Φ on $(|i\rangle + |j\rangle)/\sqrt{2}$. Since Φ is an ε-isometry, the output of Φ on this state is within trace distance 2ε of some pure state $|\gamma\rangle$. Combining Equations (7) and (10) and applying the triangle inequality yields

$$\left\| |\gamma\rangle\langle\gamma| - \frac{1}{2}\left(|\psi_i\rangle\langle\psi_i| + |\phi_i\rangle\langle\phi_j| + |\phi_j\rangle\langle\phi_i| + |\psi_j\rangle\langle\psi_j| \right) \right\|_{\text{tr}} \leq 5\varepsilon.$$

Since $|\gamma\rangle$ is pure, for some phases c_i and c_j we have

$$\left\| |\phi_i\rangle\langle\phi_j| - c_i c_j^* |\psi_i\rangle\langle\psi_j| \right\|_{\mathrm{tr}} \leq 5\varepsilon,$$

which in turn implies that

$$\left\| \Phi(|i\rangle\langle j|) - c_i c_j^* |\psi_i\rangle\langle\psi_j| \right\|_{\mathrm{tr}} \leq 9\varepsilon,$$

using Equation (10). Finally, since this is true for any $i \neq j$, and the case of $i = j$ is Equation (7), the previous equation implies that

$$\max_{\rho} \left\| \Phi(\rho) - A\rho A^* \right\|_{\mathrm{tr}} \leq 9\varepsilon,$$

where A is the operator defined by $A|i\rangle = c_i|\psi_i\rangle$ for all i.

It remains only to show that the operator $A \otimes A$ preserves symmetric states. To see this, take $|ij\rangle + |ji\rangle$ an arbitrary basis element of the symmetric subspace of $\mathcal{H}^{\otimes 2}$. By a simple calculation

$$(A \otimes A)(|ij\rangle + |ji\rangle) = c_i c_j |\psi_i\rangle \otimes |\psi_j\rangle + c_i c_j |\psi_j\rangle \otimes |\psi_i\rangle,$$

which remains invariant under swapping the two spaces. By linearity, conjugation by $A \otimes \mathbb{1}_{\mathcal{R}}$ also preserves the symmetry of states on $(\mathcal{H} \otimes \mathcal{R})^{\otimes 2}$. It follows that $\hat{\Phi}$ preserves symmetry up to an error of 9ε in the trace distance. This implies that the swap test on the output of $\hat{\Phi} \otimes \hat{\Phi}$ applied to a symmetric state returns the symmetric outcome with probability at least $1 - 9\varepsilon$.   □

This theorem shows that NON-ISOMETRY$_\varepsilon$ is in QMA for any constant ε satisfying $(1 - \varepsilon)/2 > 9\varepsilon$. Together with the QMA-hardness of the problem shown in Theorem 3 this gives the main result.

Corollary 7. *For any constant $\varepsilon < 1/19$, NON-ISOMETRY$_\varepsilon$ is QMA-complete.*

This also implies that problem of computing the channel purity, as defined by Zanardi and Lidar [20], over the whole input space is QMA-complete.

6   Conclusion

We have shown the computational intractability of the problem of detecting when a quantum channel is far from an isometry, or equivalently, when a channel can be made to output a highly mixed state. These results show that it is extremely difficult to characterize the worst-case behaviour of a quantum computation. This is similar to the classical case, where the problem of determining if a circuit can produce a specific output is known to be intractable.

We have also added to the short but growing list of problems that are known to be complete for the complexity class QMA. The NON-ISOMETRY problem provides a new way to study this class, as it exactly characterizes the difficulty of the problems in the class. It is hoped that this will lead to new results about the power of this model of computation.

There are several open problems related to this work. A few of the more interesting ones are listed below.

- As the verifier in a QMA protocol may use ancillary qubits, the hardness proof in Theorem 3 only applies to isometries. If a protocol could be constructed without these ancillary qubits, the same argument would apply to the case of testing unitarity.
- The bound of 9ε in Item 2 of Theorem 6 is hardly expected to be optimal. Improving this argument would put the problem into QMA for larger values of ε than $1/19$.
- If we take an alternate definition of approximate isometry which is that the Choi matrix is close to a maximally entangled we end up with a weaker notion of isometry. This notion can be tested in BQP using the swap test to estimate the purity of the Choi matrix, as this definition avoids the minimization of purity over all input states. Is this simpler problem complete for BQP?

Acknowledgements. I am grateful for discussions with Markus Grassl, Masahito Hayashi, Lana Sheridan, and John Watrous, from which I have learnt a great deal. This work has been supported by the Centre for Quantum Technologies, which is funded by the Singapore Ministry of Education and the Singapore National Research Foundation, as well as as well the Bell Family Fund, while the author was at the Institute for Quantum Computing at the University of Waterloo.

References

1. Aharonov, D., Kitaev, A., Nisan, N.: Quantum circuits with mixed states. In: 30th ACM Symposium on the Theory of Computing, pp. 20–30 (1998)
2. Beigi, S., Shor, P.W.: On the complexity of computing zero-error and Holevo capacity of quantum channels (2007), arXiv:0709.2090v3 [quant-ph]
3. Buhrman, H., Cleve, R., Watrous, J., de Wolf, R.: Quantum fingerprinting. Physical Review Letters 87(16), 167902 (2001)
4. Choi, M.D.: Completely positive linear maps on complex matrices. Linear Algebra and its Applications 10(3), 285–290 (1975)
5. Ekert, A.K., Alves, C.M., Oi, D.K., Horodecki, M., Horodecki, P., Kwek, L.C.: Direct estimations of linear and nonlinear functionals of a quantum state. Physical Review Letters 88(21), 217901 (2002)
6. Janzing, D., Wocjan, P., Beth, T.: "Non-identity-check" is QMA-complete. International Journal of Quantum Information 3(3), 463–473 (2005)
7. Ji, Z., Wu, X.: Non-identity check remains QMA-complete for short circuits (2009), arXiv:0906.5416 [quant-ph]
8. Kempe, J., Kitaev, A., Regev, O.: The complexity of the local Hamiltonian problem. SIAM Journal on Computing 35(5), 1070–1097 (2006)
9. Kitaev, A.Y.: Quantum NP. Talk at the 2nd Workshop on Algorithms in Quantum Information Processing (AQIP), DePaul University (1999)
10. Kitaev, A.Y., Shen, A.H., Vyalyi, M.N.: Classical and Quantum Computation. Graduate Studies in Mathematics, vol. 47. American Mathematical Society, Providence (2002)

11. Knill, E.: Quantum randomness and nondeterminism. Tech. Rep. LAUR-96-2186, Los Alamos National Laboratory (1996)
12. Liu, Y.K.: Consistency of local density matrices is QMA-complete. In: Díaz, J., Jansen, K., Rolim, J.D.P., Zwick, U. (eds.) APPROX 2006 and RANDOM 2006. LNCS, vol. 4110, pp. 438–449. Springer, Heidelberg (2006)
13. Liu, Y.K., Christandl, M., Verstraete, F.: Quantum computational complexity of the N-representability problem: QMA complete. Physical Review Letters 98(11), 110503 (2007)
14. Marriott, C., Watrous, J.: Quantum Arthur-Merlin games. Computational Complexity 14(2), 122–152 (2005)
15. Nielsen, M.A., Chuang, I.L.: Quantum Computation and Quantum Information. Cambridge University Press, Cambridge (2000)
16. Schuch, N., Cirac, I., Verstraete, F.: Computational difficulty of finding matrix product ground states. Physical Review Letters 100(25), 250501 (2008)
17. Schuch, N., Verstraete, F.: Computational complexity of interacting electrons and fundamental limitations of density functional theory. Nature Physics 5(10), 732–735 (2009)
18. Watrous, J.: Succinct quantum proofs for properties of finite groups. In: 41st IEEE Symposium on Foundations of Computer Science, pp. 537–546 (2000)
19. Wei, T.C., Mosca, M., Nayak, A.: Interacting boson problems can be QMA hard. Physical Review Letters 104(4), 040501 (2010)
20. Zanardi, P., Lidar, D.A.: Purity and state fidelity of quantum channels. Physical Review A 70(1), 012315 (2004)

Quantum Search with Advice[*]

Ashley Montanaro

Department of Applied Mathematics and Theoretical Physics,
University of Cambridge, UK
am994@cam.ac.uk

Abstract. We consider the problem of search of an unstructured list for
a marked element x, when one is given advice as to where x might be
located, in the form of a probability distribution. The goal is to minimise
the expected number of queries to the list made to find x, with respect
to this distribution. We present a quantum algorithm which solves this
problem using an optimal number of queries, up to a constant factor. For
some distributions on the input, such as certain power law distributions,
the algorithm can achieve exponential speed-ups over the best possible
classical algorithm. We also give an efficient quantum algorithm for a
variant of this task where the distribution is not known in advance, but
must be queried at an additional cost. The algorithms are based on the
use of Grover's quantum search algorithm and amplitude amplification
as subroutines.

1 Introduction

Grover's algorithm for search of an unstructured list is one of the greatest suc-
cesses of the nascent field of quantum computation [6]. The algorithm operates
in the *black box* model: given access to a function $f : \{1, \ldots, n\} \to \{0, 1\}$, where f
is promised to take the value 1 on precisely one input x, it finds x with certainty
using $O(\sqrt{n})$ queries to f, whereas any classical algorithm requires $\Omega(n)$ queries
to perform the same task. However, it is rarely necessary to search the type of
databases that we encounter in real life in a completely unstructured fashion.
Instead, there is often some prior information about the location of the sought
("marked") item x, which can be used to guide the search. We can formalise this
intuition by considering a search problem where the searcher is given access to
a probability distribution μ, which hints where the marked item is likely to be.
This problem can be stated formally as follows.

- **Problem:** SEARCH WITH ADVICE
- **Input:** A function $f : \{1, \ldots, n\} \to \{0, 1\}$ that takes the value 1 on precisely one input x, and an "advice" probability distribution $\mu = (p_y)$, $y \in$
  $\{1, \ldots, n\}$, where p_y is the probability that $f(y) = 1$.
- **Output:** The marked element x.

---

[*] Work done while at the University of Bristol.

W. van Dam et al. (Eds.): TQC 2010, LNCS 6519, pp. 77–93, 2011.
© Springer-Verlag Berlin Heidelberg 2011

It is clear that knowledge of μ can enable a classical algorithm to achieve a significant reduction in the average number of queries to f (with respect to μ) required to find the marked element x. This paper is concerned with the development of *quantum* algorithms for the SEARCH WITH ADVICE problem which also use μ, and which obtain significant speed-ups over any classical algorithm.

We distinguish two models for the complexity of this problem. In the first model – the *known* model – μ is known completely beforehand, and can be used to help design an algorithm to find the marked element. The complexity of the problem is given by the minimum expected number of queries to f required to find x, under the distribution μ. In the second model – the *unknown* model – μ is not known before the algorithm starts, but the algorithm is also given access to a black box which outputs samples from μ, at unit cost. In the case of quantum algorithms, the black box outputs a coherent superposition corresponding to μ (a "quantum sample").

In both cases, note that we are interested in the *average* number of queries with respect to μ required to find the marked element, rather than the worst-case number of queries. Previous work has shown that, if one considers the worst-case number of queries to the input required to compute any total function, there can only be at most a polynomial separation between quantum and classical computation [2]. Considering the average number of queries required (over the input) allows one to sidestep these results and hope to obtain *exponential* speed-ups.

Indeed, previous work of Ambainis and de Wolf [1] has shown that quantum algorithms can achieve exponential (or even super-exponential) reductions in average-case query complexity over classical algorithms. The model that these authors considered was that of computing a particular boolean function $f : \{0,1\}^n \rightarrow \{0,1\}$, with a particular (known) distribution on the inputs. Among other results, they exhibited a (function, distribution) pair with a super-exponential separation between quantum and classical query complexity, and even gave a function whose quantum and classical query complexity were exponentially separated under the uniform distribution.

1.1   New Results

The main results of this paper are as follows. First, in the known model, we give a quantum algorithm for SEARCH WITH ADVICE which is optimal up to constant factors. Assuming without loss of generality that the probability distribution $\mu = (p_x)$ is given in non-increasing order, the algorithm uses an expected number of queries to f which is of the order of

$$\sum_{x=1}^{n} p_x \sqrt{x},$$

which should be compared with the optimal classical expected number of queries,

$$\sum_{x=1}^{n} p_x \, x.$$

For certain probability distributions, this represents an exponential (or even super-exponential) improvement in the expected number of queries used. The quantum algorithm is based on the use of an exact variant of Grover's algorithm [6,8,4] as a subroutine. We extend known lower bounds on the query complexity of quantum search to show that this algorithm is optimal, up to constant factors, for any distribution μ.

In the unknown model, we give a quantum algorithm that uses a expected number of queries of the order of

$$\left(\sum_{x,p_x>1/n} \sqrt{p_x} \right) + \sqrt{n} \left(\sum_{x,p_x\leq 1/n} p_x \right).$$

Again, this algorithm is sometimes significantly more efficient than the best possible classical algorithm. The algorithm is based on the amplitude amplification algorithms proposed by Boyer et al [3] and Brassard et al [4]; the main difference being that after performing a certain number of iterations of amplitude amplification, it reverts to exact Grover search. This can considerably improve the average query complexity.

These results in the two different models are applied to the natural class of power law distributions $p_x \propto x^k$, for some constant $k < 0$. We will see that for certain values of k, quantum algorithms deliver very significant reductions in the average number of queries used. In particular, when $-2 < k < -3/2$, a super-exponential separation between quantum and classical computation is obtained in the known model ($O(1)$ vs. $\Omega(n^{k+2})$). The results for power law distributions are summarised in Figure 1 (with details given in Propositions 6 and 11 below).

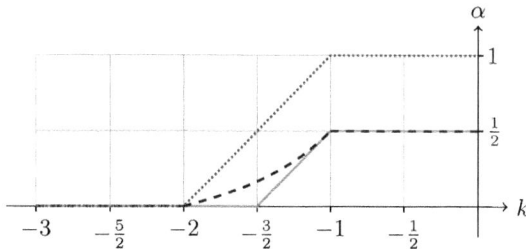

Fig. 1. Query complexity of SEARCH WITH ADVICE in different models, for power law distributions $p_x \propto x^{-k}$. For each k, the query complexity of the algorithms given in this paper is $\Theta(n^\alpha)$ for some α (ignoring log factors); the graph plots the exponent α against k. Dotted red line: best classical algorithm; solid green line: quantum, known probability distribution; dashed blue line: quantum, unknown probability distribution.

The paper is organised as follows. After defining the models and notation used, Section 2 contains the results on the known model, while Section 3 studies the unknown model. Various proofs are deferred to an appendix.

1.2   Models and Notation

In this section, we set up concepts and notation that will be used throughout the paper. We assume familiarity with quantum computation [10], and in particular the concept of query complexity [5]: the number of queries to the input which a classical or quantum algorithm requires to compute some function. Let $[n]$ denote the integers $\{1, \ldots, n\}$, and consider an oracle function $f : [n] \to \{0, 1\}$ which is promised to take the value 1 on precisely one input $x \in [n]$ (that is, $f(y) = \delta_{xy}$). We say that x is the *marked* element. Also consider a quantum or classical algorithm $\mathcal{A}$ which, given access to f, attempts to output x. We say that $\mathcal{A}$ is a *valid* algorithm if, for any f satisfying the above constraint, $\mathcal{A}$ outputs x with certainty. Let $\mathcal{D}$ denote the set of valid deterministic classical algorithms, and let $\mathcal{Q}$ denote the set of valid quantum algorithms.

Let $\mu = (p_x)$ be a distribution on $[n]$ giving the probability for the marked element to be found at each location. We will be concerned with understanding the average number of queries to the input required to find x. This will only depend on μ, and will hence be termed the (average-case) query complexity of μ. The general model of average-case query complexity used here will be similar to that in [1], with some minor differences. We distinguish two models for the SEARCH WITH ADVICE problem: a *known* model and an *unknown* model. In the known model, the probability distribution μ is known beforehand, and can be used to design the algorithm. In the unknown model, μ is not known, and the algorithm must query an oracle to gain information about μ.

We first define the known model. Let $\mathcal{A}$ be a valid algorithm, and let $T_{\mathcal{A}}(x)$ denote the expected number of queries to f used by $\mathcal{A}$, when x is the marked element. Note that, in order for this model to be interesting in the case where $\mathcal{A}$ is a quantum algorithm, intermediate measurements during the search process are allowed; otherwise, $\mathcal{A}$ would always use the same number of queries. Further, let $T_{\mathcal{A}}(\mu)$ be the expected number of queries to f used by $\mathcal{A}$, where the expectation is taken over both the distribution μ and (potentially) $\mathcal{A}$'s internal randomness. That is,

$$T_{\mathcal{A}}(\mu) = \sum_{x=1}^{n} p_x T_{\mathcal{A}}(x).$$

Finally, we define the main quantities of interest, the deterministic and quantum (respectively) average-case query complexities of μ.

$$D(\mu) = \min_{\mathcal{A} \in \mathcal{D}} T_{\mathcal{A}}(\mu),$$
$$Q(\mu) = \min_{\mathcal{A} \in \mathcal{Q}} T_{\mathcal{A}}(\mu).$$

The restriction to algorithms that succeed with certainty makes this a zero-error (*Las Vegas*) notion of average-case query complexity. It is common to consider an alternative *Monte Carlo* model of query complexity where $\mathcal{A}$ is allowed to err with some constant probability (e.g. $1/3$). Note that this would not change the model significantly in the case of the current problem: given a (classical or quantum) Monte Carlo search algorithm that uses t queries and outputs x with probability p, one can produce an algorithm that succeeds with certainty and uses an expected number of queries of at most $(t+1)/p$ [9, Exercise 1.3].

We now turn to the unknown model. In this scenario, as well as querying f, we allow $\mathcal{A}$ to sample from μ using an oracle. In the case of quantum algorithms, we allow the preparation of *quantum* samples; that is, in the quantum case we define an oracle O_μ, which performs the mapping

$$O_\mu|0\rangle = |\mu\rangle := \sum_{x=1}^{n} \sqrt{p_x}|x\rangle.$$

The algorithm is also given access to the inverse operation, O_μ^{-1}. We define $T_\mathcal{A}^*(\mu)$ as the expected total number of queries to f, O_μ and O_μ^{-1} used by $\mathcal{A}$ (a query to each oracle being counted as unit cost). The oracle O_μ may appear somewhat unrealistic. However, it can be implemented if one has the ability to sum the distribution μ over arbitrary ranges [7]. That is, given an efficient means of computing $\sum_{x=a}^{b} p_x$ for arbitrary a, b, one can implement O_μ efficiently.

Finally, we will make use of an exact variant of Grover's quantum search algorithm throughout this paper.

Theorem 1 (Grover [6], Høyer [8], Brassard et al [4]). *Given an unstructured list of n elements that contains a unique marked element, there is a quantum algorithm that finds the marked element with certainty using $\lceil \frac{\pi}{4}\sqrt{n} \rceil$ queries to the list. If the list is promised to contain either one or zero marked elements, the marked element can be found (or "no marked element present" returned) with certainty using one extra query.*

2   Search with a Known Probability Distribution

In this section, we will assume that p_x is non-increasing with x (so the most likely place for the marked element to be is at the start of the list, etc.). With this assumption, the optimal classical algorithm to find x is simply to query $f(1)$ through $f(n)$ in turn, so the classical average-case query complexity can be written down as

$$D(\mu) = \sum_{x=1}^{n} p_x \, x. \tag{1}$$

Note that, classically, the algorithm obtains no benefit from the use of randomness. When μ is the uniform distribution, corresponding to having no information about the location of the marked item, Grover's algorithm (Theorem 1) achieves a quadratic reduction in average-case query complexity. However, naïve use of this algorithm does not give an advantage in the average-case setting in general. In the next section, we give a quantum algorithm which does significantly improve on the trivial classical algorithm above.

2.1 Geometric Search Algorithm

We give a general algorithm for the SEARCH WITH ADVICE problem, which will turn out to be asymptotically optimal. The quantum component of this algorithm is in fact simply Grover search (Theorem 1). Informally, the algorithm consists of splitting the input into blocks which increase in size geometrically (hence its name) and performing Grover search on each block. Interestingly, the algorithm does not need to know the precise advice probability distribution to achieve its near-optimal query complexity; it suffices to be able to sort the probabilities in non-increasing order. The algorithm is parametrised by a constant c, which gives the ratio of the geometric progression. We optimise c below; however, changing c only affects the query complexity by a constant factor.

Algorithm 1. Geometric quantum search

Input: Advice distribution $\mu = (p_x)$ in non-increasing order; function
      $f : [n] \to \{0, 1\}$ such that f takes the value 1 on precisely one input x;
      real $c > 1$
Output: The marked element x
$start \leftarrow 1$;
$end \leftarrow 1$;
$step \leftarrow 0$;
while $start \leq n$ **do**
    perform exact Grover search for one or zero marked elements on subset
    $\{start, \ldots, end\}$;
    if *marked element found* **then**
        return *marked element*;
    end
    $step \leftarrow step + 1$;
    $start \leftarrow end + 1$;
    $end \leftarrow \min(start + \lfloor c^{step} \rfloor - 1, n)$;
end
return *error*;

Proposition 2. *The average number of queries used by Algorithm 1, choosing $c = e \approx 2.718$, on an advice distribution $\mu = (p_x)$ is upper bounded by*

$$\pi e \sum_{x=1}^{n} p_x \sqrt{x}.$$

Proof. In the m'th iteration of the loop, the (at most) $\lfloor c^m \rfloor$ elements contained in the range

$$R_m = \{1 + \sum_{i=0}^{m-1} \lfloor c^i \rfloor, \ldots, \min(\lfloor c^m \rfloor - \sum_{i=0}^{m-1} \lfloor c^i \rfloor, n)\} \tag{2}$$

will be searched. By Theorem 1, the Grover search step in this iteration uses $\left\lceil \frac{\pi}{4} \sqrt{\lfloor c^m \rfloor} \right\rceil + 1$ queries. Then, for any marked element $x \in R_m$, a total of at most

$$\sum_{s=0}^{m} \left(\left\lceil \frac{\pi}{4} \sqrt{\lfloor c^s \rfloor} \right\rceil + 1 \right) \leq 2(m+1) + \frac{\pi}{4} \sum_{s=0}^{m} c^{s/2}$$

queries will be used by Algorithm 1 to find x. It is clear from (2) that, for any $x \in R_m$, $m \leq \log_c x + 1$. The average-case query complexity is therefore upper bounded by

$$\sum_{x=1}^{n} p_x \left(2 \log_c x + 4 + \frac{\pi}{4} \sum_{s=0}^{\lfloor \log_c x + 1 \rfloor} c^{s/2} \right),$$

and, estimating the inner sum by an integral, we obtain an upper bound of

$$\sum_{x=1}^{n} p_x (2 \log_c x + 4 + \frac{\pi}{4} \int_0^{\log_c x + 2} c^{s/2} \, ds) = 2 \sum_{x=1}^{n} p_x (\log_c x + 2 + \frac{\pi c}{4 \ln c} \sqrt{x}).$$

Picking $c = e$, and noting that $\ln x + 2 \leq \frac{\pi e}{4} \sqrt{x}$ for all $x > 0$, completes the proof. $\square$

2.2 Optimality of the Geometric Search Algorithm

We now show that Algorithm 1 is in fact optimal, up to a constant factor. This result will rely on the following known exact bound on the query complexity of quantum search.

Theorem 3 (Grover [6], Zalka [12]). *Let $f : [n] \rightarrow \{0,1\}$ be a function that takes the value 1 on precisely one input x, and let A be a quantum search algorithm that uses T queries to f and outputs x with probability at least p, for all x. Then*

$$T \geq \left\lceil \frac{\arcsin \sqrt{p}}{2 \arcsin(1/\sqrt{n})} - \frac{1}{2} \right\rceil,$$

and this number of queries is achieved by Grover's algorithm.

As stated, this bound involves worst-case query complexity (that is, the largest possible number of queries used by A, on the worst possible input). In our setting, we will need to lower bound the *expected* number of queries used by $\mathcal{A}$ on the worst possible input. This can be done with the following proposition.

Proposition 4. *Let $\mathcal{A}$ be a valid quantum search algorithm such that $T_{\mathcal{A}}(x) \leq T$ for all x, for some T. Then*

$$T \geq \frac{0.206}{\arcsin 1/\sqrt{n}} - 0.316 \geq 0.206\sqrt{n} - 1.$$

Proof. Let $t_{\mathcal{A}}(x)$ be the random variable giving the number of queries used by $\mathcal{A}$ on input x. Thus $T_{\mathcal{A}}(x) = \mathbb{E}\, t_{\mathcal{A}}(x)$, where the expectation is taken over $\mathcal{A}$'s internal randomness. By Markov's inequality, for all x and all $0 < p < 1$,

$$\Pr[t_{\mathcal{A}}(x) \geq T_{\mathcal{A}}(x)/(1-p)] \leq (1-p).$$

Thus a quantum search algorithm $\mathcal{A}$ that uses an expected number of at most T queries on all x gives a bounded-error quantum search algorithm that uses at most $T/(1-p)$ queries on all x and succeeds with probability at least p: just run $\mathcal{A}$ until it has used $T/(1-p)$ queries, and if it has not output x, output a random integer between 1 and n. By Markov's inequality, this will succeed with probability at least p.

So, by Theorem 3, we have that for any $0 < p < 1$

$$T \geq (1-p) \left(\frac{\arcsin\sqrt{p}}{2\arcsin(1/\sqrt{n})} - \frac{1}{2} \right).$$

Performing numerical maximisation of the right-hand side over p, one finds that for large n the maximum is achieved at $p \approx 0.369$, which proves the proposition. $\square$

Note that it is known that one can indeed achieve an expected query complexity that is somewhat less than the usual worst-case query complexity guaranteed by Grover's algorithm [3,12]. By stopping and restarting Grover search, it is possible to find the marked element x using approximately $0.690\sqrt{n}$ expected queries on all x, whereas straightforward use of Grover's algorithm guarantees approximately $0.785\sqrt{n}$ queries.

We are now ready to prove that Algorithm 1 is asymptotically optimal.

Proposition 5. *Let $\mu = (p_x)$, $x \in [n]$ be an arbitrary probability distribution. Then*

$$Q(\mu) \geq 0.206 \sum_{x=1}^{n} p_x \sqrt{x} - 1.$$

Proof. Let $\mathcal{A}$ be a valid quantum search algorithm and assume that μ is non-increasing. We aim to lower bound $T_{\mathcal{A}}(\mu) = \sum_{x=1}^{n} p_x T_{\mathcal{A}}(x)$. By Proposition 4, there must exist a y such that $T_{\mathcal{A}}(y) \geq 0.206\sqrt{n} - 1$. Similarly, there must exist $y' \neq y$ such that $T_{\mathcal{A}}(y') \geq 0.206\sqrt{n-1} - 1$ (or $\mathcal{A}$ would be able to find a marked element in the set of all elements not equal to y, using a number of queries that violates Proposition 4). Iterating this argument, we see that for each k such that $1 \leq k \leq n$ there exists an x such that $T_A(x) \geq 0.206\sqrt{k} - 1$. By a rearrangement inequality, this implies that

$$T_A(\mu) \geq \sum_{x=1}^{n} p_x(0.206\sqrt{x} - 1)$$

and proves the proposition.   □

2.3 Power Law Distributions

We now apply Algorithm 1 to a natural class of probability distributions: power law distributions. We will see that significant speed-ups can be obtained over any possible classical algorithm.

Proposition 6. Let $\mu = (p_x)$, $x \in [n]$ be a probability distribution where $p_x \propto x^k$ for some constant $k < 0$. Then

$$D(\mu) = \begin{cases} \Theta(n) & [-1<k<0] \\ \Theta(n/\log n) & [k=-1] \\ \Theta(n^{k+2}) & [-2<k<-1] \\ \Theta(\log n) & [k=-2] \\ \Theta(1) & [k<-2] \end{cases}, \quad Q(\mu) = \begin{cases} \Theta(\sqrt{n}) & [-1<k<0] \\ \Theta(\sqrt{n}/\log n) & [k=-1] \\ \Theta(n^{k+3/2}) & [-3/2<k<-1] \\ \Theta(\log n) & [k=-3/2] \\ \Theta(1) & [k<-3/2] \end{cases}$$

Proof. Deferred to Appendix.

Corollary 7. *There exists a probability distribution μ such that $D(\mu) = \Omega(n^{1/2-\epsilon})$ for arbitrary $\epsilon > 0$, but $Q(\mu) = O(1)$.*

Proof. Take $k = -3/2 - \epsilon$ in Proposition 6. (Indeed, any $k \in (-2, -3/2)$ gives a super-exponential separation between $D(\mu)$ and $Q(\mu)$.)   □

3 Unknown Probability Distribution

In this section we switch to a different model, where the algorithm does not know the advice distribution μ in advance, but must use an oracle to obtain information about this distribution. We begin by noting the somewhat counterintuitive fact that a classical algorithm that merely queries f according to samples from the distribution μ performs no better than an exhaustive search algorithm[1].

Indeed, consider a classical algorithm that consists of repeatedly obtaining a sample y from μ, then querying $f(y)$. If x is the marked element, the expected number of samples from μ required until x is found is exactly $1/p_x$ (assuming that $p_x > 0$). Thus the expected number of samples used is

$$\sum_{x=1}^{n} p_x\left(\frac{1}{p_x}\right) = n;$$

---

[1] This phenomenon was recently discussed in the somewhat different context of screening for terrorists [11].

the algorithm might as well have just carried out an exhaustive search to find x. Being given access to a *quantum* oracle O_μ producing a coherent superposition corresponding to the distribution μ, however, will turn out to be very useful.

3.1 Quantum Algorithm

Our quantum algorithm will be based on the *amplitude amplification* primitive of Brassard et al [4]. For completeness, an explicit definition of amplitude amplification is given below, as Algorithm 2.

Algorithm 2. Amplitude amplification [4]

Input: Function $f : [n] \rightarrow \{0, 1\}$ such that f takes the value 1 on precisely one
    input x; oracle operator $O_\mu : |0\rangle \mapsto |\mu\rangle$; inverse O_μ^{-1}; positive integer i
    (number of iterations)
Output: The marked element x, or fail
create initial state $|\mu\rangle = O_\mu|0\rangle$;
apply operator $-O_\mu I_{|0\rangle} O_\mu^{-1} I_{|x\rangle}$ i times to $|\mu\rangle$;
measure in computational basis, obtaining outcome y;
if $f(y)=1$ **then**
    return y;
else
    return *fail*;
end

The notation $I_{|\psi\rangle}$ denotes reflection about the state $|\psi\rangle$; it is well-known that the operator $I_{|x\rangle}$, where $I_{|x\rangle}|y\rangle = (-1)^{f(y)}|y\rangle$, can be implemented using one query to f. The following result was shown by Brassard et al in [4] (with somewhat different terminology).

Lemma 8. *Applying Algorithm 2 with i iterations returns the location of the marked element with probability* $\sin^2((2i + 1) \arcsin \sqrt{p_x})$, *using $i + 1$ queries to O_μ, i queries to O_μ^{-1}, and $i + 1$ queries to f.*

We now use Algorithm 2 as a subroutine in an algorithm which finds the marked element with certainty, and takes advantage of O_μ to reduce the expected number of queries used. The algorithm is a modified version of previous "exponential searching" algorithms of Brassard et al [4], and Boyer et al [3]. The main difference is that the algorithm gives up after a certain number of iterations and reverts to the exact variant of standard Grover search [8,4]. This change can make a significant difference to the overall query complexity. The algorithm is stated as Algorithm 3 below.

Algorithm 3. Quantum search with unknown probability distribution

---

Input: Function $f : [n] \to \{0, 1\}$ such that f takes the value 1 on precisely one
input x; oracle operator $O_\mu : |0\rangle \mapsto |\mu\rangle$; inverse O_μ^{-1}; real $c > 1$
Output: The marked element x
for $j = 0$ *to* $\lfloor \log_c \sqrt{n} \rfloor$ **do**
    sample from distribution μ;
    if *marked element found* **then**
        return *marked element*;
    end
    pick i uniformly at random from integers $\{0, \dots, \lfloor c^j \rfloor - 1\}$;
    perform i iterations of amplitude amplification;
    if *marked element found* **then**
        return *marked element*;
    end
end
perform exact Grover search for one marked element on $[n]$;
return *marked element*;

It will turn out to be possible to give a close analysis of the expected query complexity of Algorithm 3, including constants (which we will round to integers; these could be optimised further). The analysis follows the approach taken by Boyer et al [3] to bound the performance of their quantum search algorithm for an unknown number of marked elements.

Proposition 9. *On input x, when called with $c \approx 1.162$, Algorithm 3 uses an expected number of at most $\min\{83/\sqrt{p_x} + 4/3, 53\sqrt{n}\}$ queries to each of f, O_μ, O_μ^{-1}.*

Proof. Deferred to Appendix.

Corollary 10. *Let $\mathcal{A}$ denote Algorithm 3. Then there are constants K, L, M such that*

$$T_{\mathcal{A}}^*(\mu) \leq K \left(\sum_{x, p_x > 1/n} \sqrt{p_x} \right) + L\sqrt{n} \left(\sum_{x, p_x \leq 1/n} p_x \right) + M.$$

Proof. Apply Algorithm 3, and use Proposition 9 to calculate the average number of queries used with respect to the distribution μ. □

3.2 Power Law Distributions

As with the case of a known probability distribution, power law distributions provide a natural class of examples for search with an unknown probability distribution. For some of these distributions, Algorithm 3 can be used to obtain significant speed-ups over any classical algorithm, even one with complete knowledge of the distribution.

Proposition 11. *Let $\mu = (p_x)$ be a probability distribution where $p_x \propto x^k$ for some constant $k < 0$, and let $\mathcal{A}$ denote Algorithm 3. Then*

$$
T_{\mathcal{A}}^*(\mu) = \begin{cases}
O(\sqrt{n}) & [-1 \le k < 0] \\
O(n^{-(1/2+1/k)}) & [-2 < k < -1] \\
O(\log n) & [k = -2] \\
O(1) & [k < -2]
\end{cases}
$$

Proof. Deferred to Appendix.

Acknowledgements

This work was supported by the EC-FP6-STREP network QICS and an EPSRC Postdoctoral Research Fellowship. I would like to thank Aram Harrow for pointing out reference [11].

References

1. Ambainis, A., de Wolf, R.: Average-case quantum query complexity. J. Phys. A: Math. Gen. 34, 6741–6754 (2001), quant-ph/9904079
2. Beals, R., Buhrman, H., Cleve, R., Mosca, M., de Wolf, R.: Quantum lower bounds by polynomials. J. ACM 48(4), 778–797 (2001), quant-ph/9802049
3. Boyer, M., Brassard, G., Høyer, P., Tapp, A.: Tight bounds on quantum searching. Fortschr. Phys. 46(4-5), 493–505 (1998), quant-ph/9605034
4. Brassard, G., Høyer, P., Mosca, M., Tapp, A.: Quantum amplitude amplification and estimation. In: Quantum Computation and Quantum Information: A Millennium Volume, pp. 53–74 (2002), quant-ph/0005055
5. Buhrman, H., de Wolf, R.: Complexity measures and decision tree complexity: a survey. Theoretical Computer Science 288, 21–43 (2002)
6. Grover, L.: Quantum mechanics helps in searching for a needle in a haystack. Phys. Rev. Lett. 79(2), 325–328 (1997), quant-ph/9706033
7. Grover, L., Rudolph, T.: Creating superpositions that correspond to efficiently integrable probability distributions (2002), quant-ph/0208112
8. Høyer, P.: Arbitrary phases in quantum amplitude amplification. Phys. Rev. A 62, 052304 (2000), quant-ph/0006031
9. Motwani, R., Raghavan, P.: Randomized algorithms. Cambridge University Press, Cambridge (1995)
10. Nielsen, M.A., Chuang, I.L.: Quantum computation and quantum information. Cambridge University Press, Cambridge (2000)
11. Press, W.H.: Strong profiling is not mathematically optimal for discovering rare malfeasors. Proceedings of the National Academy of Sciences 106(6), 1716–1719 (2009)
12. Zalka, C.: Grover's quantum searching algorithm is optimal. Phys. Rev. A. 60(4), 2746–2751 (1999), quant-ph/9711070

Appendix

In this appendix we collect some proofs from throughout the paper.

A.1   Proofs from Section 2

Proposition 6. *Let* $\mu = (p_x)$, $x \in [n]$ *be a probability distribution where* $p_x \propto x^k$ *for some constant* $k < 0$. *Then*

$$D(\mu) = \begin{cases} \Theta(n) & [-1<k<0] \\ \Theta(n/\log n) & [k=-1] \\ \Theta(n^{k+2}) & [-2<k<-1] \\ \Theta(\log n) & [k=-2] \\ \Theta(1) & [k<-2] \end{cases}, Q(\mu) = \begin{cases} \Theta(\sqrt{n}) & [-1<k<0] \\ \Theta(\sqrt{n}/\log n) & [k=-1] \\ \Theta(n^{k+3/2}) & [-3/2<k<-1] \\ \Theta(\log n) & [k=-3/2] \\ \Theta(1) & [k<-3/2] \end{cases}$$

Proof. From the statement of the proposition, $p_x = \alpha x^k$ for some α. We first estimate the normalising constant α. The constraint that $\sum_{x=1}^{n} p_x = 1$ implies that $1/\alpha = \sum_{x=1}^{n} x^k$. Estimating this sum by an integral, we have that

$$\int_1^n x^k \, dx \leq 1/\alpha \leq 1 + \int_1^n x^k \, dx,$$

implying that, for $k \neq -1$,

$$\frac{n^{k+1} - 1}{k + 1} \leq 1/\alpha \leq \frac{n^{k+1} - 1}{k + 1} + 1.$$

By (1), if it also holds that $k \neq -2$,

$$D(\mu) = \alpha \sum_{x=1}^{n} x^{k+1} \geq \alpha \int_1^n x^{k+1} \, dx \geq \frac{(n^{k+2} - 1)(k + 1)}{(n^{k+1} + k)(k + 2)}.$$

The upper bound on $D(\mu)$ is very similar. It is easy to see that this proves the deterministic half of the proposition, except for the cases $k = -1$ and $k = -2$, which can be verified directly. In the quantum case, by Proposition 2, for $k \neq -1, -3/2$,

$$Q(\mu) \leq \alpha \pi e \sum_{x=1}^{n} x^{k+1/2} \leq \alpha \pi e \left(1 + \int_1^{n+1} x^{k+1/2} \, dx \right)$$

$$= \pi e \left(\frac{((n + 1)^{k+3/2} + k + 1/2)(k + 1)}{(n^{k+1} - 1)(k + 3/2)} \right).$$

Again, the lower bound is similar and the special cases $k = -1$, $k = -3/2$ can be verified directly. $\qquad\square$

A.2   Proofs from Section 3

In this section, we give the proof of Proposition 9, which bounds the performance of Algorithm 3. We will need a lemma of Boyer et al [3], which we translate into our terminology.

Lemma A.1 (Boyer et al [3]). *If an integer r is picked from the range $\{0, \ldots, m-1\}$ uniformly at random, and r iterations of amplitude amplification are performed, the probability of finding the marked element is exactly*

$$P_m = \frac{1}{2} - \frac{\sin(4m \arcsin \sqrt{p_x})}{8m\sqrt{p_x(1 - p_x)}}.$$

In particular, $P_m \geq 1/4$ whenever

$$m \geq \frac{1}{2\sqrt{p_x(1 - p_x)}}.$$

We are now ready to prove Proposition 9. The proof is similar to a result of Boyer et al [3], but with somewhat more detail.

Proposition 9. *On input x, when called with $c \approx 1.162$, Algorithm 3 uses an expected number of at most $\min\{83/\sqrt{p_x} + 4/3, 53\sqrt{n}\}$ queries to each of f, O_μ, O_μ^{-1}.*

Proof. We upper bound the expected number of queries to f used by Algorithm 3, which implies the same bound on the number of queries to O_μ and O_μ^{-1}. The bound will be in terms of c, and eventually minimised over c such that $1 < c < 4/3$. However, changing c will only change the number of queries used by a constant factor. Let T_j denote the expected number of queries to f used if the marked element is found in the j'th iteration of the loop. Then

$$T_j = \sum_{i=0}^{j-1} \frac{\lfloor c^i \rfloor + 3}{2} \leq \frac{c^j}{c - 1},$$

an inequality which holds for $1 < c < 4/3$ and can be proven by induction on j.

   We first find an upper bound by considering the *worst-case* number of queries used. If it has not been found previously, the marked element is guaranteed to be found in the last, exact Grover search step. Thus the number of queries to f used is at most

$$\sum_{j=1}^{\lfloor \log_c \sqrt{n} \rfloor + 1} T_j + \left\lceil \frac{\pi}{4}\sqrt{n} \right\rceil \leq \frac{1}{c-1} \sum_{j=1}^{\lfloor \log_c \sqrt{n} \rfloor + 1} c^j + \left\lceil \frac{\pi}{4}\sqrt{n} \right\rceil$$

$$\leq \left(\frac{c^2}{(c-1)^2} + \frac{\pi}{4} \right) \sqrt{n}.$$

This deals with one half of the statement of the proposition. For the remainder of the proof, we restrict to the case $p_x \geq 1/n$ (as the case $p_x \leq 1/n$ will be covered by the above bound), and also assume that $p_x \leq 3/4$; this assumption will be removed at the end.

We now assume that the "sample from distribution μ" step always fails (this can only increase the number of queries used). The expected number of queries to f used by Algorithm 3 is then upper bounded by

$$\sum_{j=0}^{\lfloor \log_c \sqrt{n} \rfloor} \left(\prod_{i=0}^{j-1} (1 - P_{\lfloor c^i \rfloor}) \right) P_{\lfloor c^j \rfloor} T_{j+1}$$

$$+ \left(\prod_{i=0}^{\lfloor \log_c \sqrt{n} \rfloor} (1 - P_{\lfloor c^i \rfloor}) \right) \left(T_{\lfloor \log_c \sqrt{n} \rfloor + 1} + \left\lceil \frac{\pi}{4} \sqrt{n} \right\rceil \right).$$

To bound this expression, we split the first sum into two parts. First, we have

$$\sum_{j=0}^{\lfloor \log_c 1/\sqrt{p_x} \rfloor} \left(\prod_{i=0}^{j-1} (1 - P_{\lfloor c^i \rfloor}) \right) P_{\lfloor c^j \rfloor} T_{j+1} \leq T_{\lfloor \log_c 1/\sqrt{p_x} \rfloor + 1} \leq \frac{c}{c-1} \frac{1}{\sqrt{p_x}}.$$

Using Lemma A.1 and the fact that $p_x \leq 3/4$, it holds for all j such that $j \geq \lfloor \log_c 1/\sqrt{p_x} \rfloor + 1$ that

$$\prod_{i=0}^{j} (1 - P_{\lfloor c^i \rfloor}) \leq \left(\frac{3}{4} \right)^{j - \lfloor \log_c 1/\sqrt{p_x} \rfloor}.$$

Thus

$$\sum_{j=\lfloor \log_c 1/\sqrt{p_x} \rfloor + 1}^{\lfloor \log_c \sqrt{n} \rfloor} \left(\prod_{i=0}^{j-1} (1 - P_{\lfloor c^i \rfloor}) \right) P_{\lfloor c^j \rfloor} T_{j+1}$$

$$\leq \frac{c}{c-1} \sum_{j=\lfloor \log_c 1/\sqrt{p_x} \rfloor + 1}^{\lfloor \log_c \sqrt{n} \rfloor} \left(\frac{3}{4} \right)^{j - \lfloor \log_c 1/\sqrt{p_x} \rfloor - 1} c^j$$

$$\leq \frac{c}{c-1} \left(\frac{4}{3} \right)^{\lfloor \log_c 1/\sqrt{p_x} \rfloor + 1} \sum_{j=0}^{\infty} \left(\frac{3c}{4} \right)^{j + \lfloor \log_c 1/\sqrt{p_x} \rfloor + 1}$$

$$\leq \frac{4c^2}{(c-1)(4-3c)} \frac{1}{\sqrt{p_x}},$$

and also

$$\left(\prod_{i=0}^{\lfloor \log_c \sqrt{n} \rfloor} (1 - P_{\lfloor c^i \rfloor}) \right) \left(T_{\lfloor \log_c \sqrt{n} \rfloor + 1} + \left\lceil \frac{\pi}{4} \sqrt{n} \right\rceil \right)$$

$$\leq \left(\frac{3}{4} \right)^{\lfloor \log_c \sqrt{n} \rfloor - \lfloor \log_c 1/\sqrt{p_x} \rfloor} \left(T_{\lfloor \log_c \sqrt{n} \rfloor + 1} + \left\lceil \frac{\pi}{4} \sqrt{n} \right\rceil \right)$$

$$\leq \left(\frac{3}{4} \right)^{\log_c \sqrt{n} - \log_c 1/\sqrt{p_x} - 1} \left(\left(\frac{c}{c-1} + \frac{\pi}{4} \right) \sqrt{n} + 1 \right)$$

$$\leq \left(\frac{4c}{3(c-1)} + \frac{\pi}{3} \right) \frac{1}{\sqrt{p_x}} + \frac{4}{3},$$

where we use again the restriction that $p_x \geq 1/n$. Combining these bounds gives the following overall upper bound on the expected number of queries used:

$$\left(\frac{c}{c-1} + \frac{4c^2}{(c-1)(4-3c)} + \frac{4c}{3(c-1)} + \frac{\pi}{3}\right)\frac{1}{\sqrt{p_x}} + \frac{4}{3}.$$

Minimising the bracketed expression over c using simple calculus gives that the minimum is found at $c \approx 1.162$; for this value of c, we obtain a bound on the expected number of queries used that is approximately

$$\frac{82.646}{\sqrt{p_x}} + \frac{4}{3}.$$

Finally, consider the case that $p_x \geq 3/4$. In this case, one can find a bound by assuming that only the sampling step in each iteration of the loop can succeed, and ignoring the amplitude amplification step. Using this assumption, the number of queries to f used is upper bounded by

$$\frac{3}{4}\sum_{j=0}^{\lfloor \log_c \sqrt{n} \rfloor}\left(\frac{1}{4}\right)^j(1+T_j) + \left(\frac{1}{4}\right)^{\lfloor \log_c \sqrt{n} \rfloor + 1}\left(T_{\lfloor \log_c \sqrt{n} \rfloor + 1} + \left\lceil\frac{\pi}{4}\sqrt{n}\right\rceil\right),$$

which is readily seen to be upper bounded by

$$\frac{3}{(c-1)(4-c)} + \frac{c}{c-1} + 2 + \frac{\pi}{4}.$$

Inserting the previously found value of c, $c \approx 1.162$, gives an upper bound of an expected ≈ 16.500 queries used in this case, and completes the proof.   □

Proposition 11. *Let* $\mu = (p_x)$ *be a probability distribution where* $p_x \propto x^k$ *for some constant* $k < 0$, *and let* $\mathcal{A}$ *denote Algorithm 3. Then*

$$T_{\mathcal{A}}^*(\mu) = \begin{cases} O(\sqrt{n}) & [-1 \leq k < 0] \\ O(n^{-(1/2+1/k)}) & [-2 < k < -1] \\ O(\log n) & [k = -2] \\ O(1) & [k < -2] \end{cases}$$

Proof. As in the proof of Proposition 6, $p_x = \alpha x^k$ for some α, and for $k \neq -1$,

$$\alpha \leq \frac{k+1}{n^{k+1} - 1}.$$

Define $x_0 = \max\{x \in [n] : p_x \geq 1/n\} = \lfloor(\alpha n)^{-1/k}\rfloor$. Then, by Corollary 10,

$$T_{\mathcal{A}}^*(\mu) \leq K\sqrt{\alpha}\sum_{x=1}^{x_0}x^{k/2} + L\alpha\sqrt{n}\sum_{x=x_0+1}^{n}x^k + M,$$

for some constants K, L, M, implying that for $k \neq -1$, $k \neq -2$,

$$
\begin{aligned}
T_{\mathcal{A}}^*(\mu) &\leq K\sqrt{\alpha}\left(1 + \int_1^{x_0} x^{k/2}\,dx\right) + L\,\alpha\sqrt{n}\int_{x_0}^n x^k\,dx + M \\
&\leq K\sqrt{\alpha}\left(1 - \frac{2}{k+2} + \frac{2x_0^{k/2+1}}{k+2}\right) + L\frac{\alpha\sqrt{n}}{k+1}(n^{k+1} - x_0^{k+1}) + M \\
&\leq K\sqrt{\frac{k+1}{n^{k+1}-1}}\left(1 + \frac{2x_0^{k/2+1} - 2}{k+2}\right) + L\frac{\sqrt{n}(n^{k+1} - x_0^{k+1})}{n^{k+1}-1} + M.
\end{aligned}
$$

Now note that

$$
x_0 \approx \left(\frac{n(k+1)}{n^{k+1}-1}\right)^{-1/k}
$$

is $\Theta(n)$ for $k > -1$, and $\Theta(n^{-1/k})$ for $k < -1$. Inserting this into the previous expression we obtain the claimed results for the cases $k \neq -1$, $k \neq -2$. These remaining special cases can be verified directly.   □

Simulating Sparse Hamiltonians
with Star Decompositions[*]

Andrew M. Childs[1,3] and Robin Kothari[2,3]

[1] Department of Combinatorics & Optimization, University of Waterloo
[2] David R. Cheriton School of Computer Science, University of Waterloo
[3] Institute for Quantum Computing, University of Waterloo

Abstract. We present an efficient algorithm for simulating the time evolution due to a sparse Hamiltonian. In terms of the maximum degree d and dimension N of the space on which the Hamiltonian H acts for time t, this algorithm uses $(d^2(d+\log^* N)\,\|Ht\|)^{1+o(1)}$ queries. This improves the complexity of the sparse Hamiltonian simulation algorithm of Berry, Ahokas, Cleve, and Sanders, which scales like $(d^4(\log^* N)\,\|Ht\|)^{1+o(1)}$. To achieve this, we decompose a general sparse Hamiltonian into a small sum of Hamiltonians whose graphs of non-zero entries have the property that every connected component is a star, and efficiently simulate each of these pieces.

1 Introduction

Quantum simulation of Hamiltonian dynamics is a well-studied problem [1,2,3] and is one of the main motivations for building a quantum computer. Since the best known classical algorithms for simulating quantum dynamics are inefficient, Feynman suggested that computers that are inherently quantum might be better at simulating quantum systems [4]. Besides simulating physics, Hamiltonian simulation has algorithmic applications, such as adiabatic optimization [5], unstructured search [6], and the implementation of continuous-time quantum walks [7,8].

The input to the Hamiltonian simulation problem is a Hamiltonian H and a time t; the problem is to implement the unitary operator e^{-iHt} approximately. We say that a Hamiltonian acting on an N-dimensional quantum system can be simulated efficiently if there is a quantum circuit using $\mathrm{poly}(\log N, t, 1/\epsilon)$ one- and two-qubit gates that approximates (with error at most ϵ) the evolution according to H for time t. Since the time evolution depends on the product Ht, the size of the circuit should also be bounded by a polynomial in some quantity measuring the size of H. When H is sparse, most of its matrix norms have comparable values, so the complexity of simulating H is not very sensitive to how its size is quantified. It is conventional to require that the scaling be polynomial in $\|H\|$, the spectral norm of H.

---

[*] Work supported by MITACS, NSERC, QuantumWorks, and the US ARO/DTO.

W. van Dam et al. (Eds.): TQC 2010, LNCS 6519, pp. 94–103, 2011.

Lloyd presented a method for simulating quantum systems that can be described by a sum of local Hamiltonians [1]. A Hamiltonian is called local if it acts non-trivially on at most a fixed number of qubits, independent of the size of the system.

This was later generalized by Aharonov and Ta-Shma [2] to the case of sparse (and efficiently row-computable) Hamiltonians. A Hamiltonian is sparse if it has at most poly($\log N$) nonzero entries in any row. It is efficiently row-computable if there is an efficient procedure to determine the location and matrix elements of the nonzero entries in each row.

The complexity of this simulation was improved by Childs [9] and further improved by Berry, Ahokas, Cleve and Sanders [3]. Their algorithm has query complexity $(d^4(\log^* N) \|Ht\|)^{1+o(1)}$, where d is the maximum degree of the graph of the Hamiltonian H. These algorithms decompose the Hamiltonian into a sum of Hamiltonians, each of which is easy to simulate. In this paper, we present a different method of decomposing the Hamiltonian, giving an algorithm with query complexity $(d^2(d + \log^* N) \|Ht\|)^{1+o(1)}$.

Note that the simulation of Ref. [3] has also been improved using a completely different approach [10,11]. That algorithm is more efficient in terms of all parameters except the error ϵ, on which its dependence is considerably worse. The algorithm we present here maintains the same dependence on ϵ as in Ref. [3], providing the best known method for high-precision simulation of sparse Hamiltonians.

2   Hamiltonians and Graphs

A Hamiltonian H acting on n qubits is a $2^n \times 2^n$ Hermitian matrix. It can also be thought of as the weighted adjacency matrix of a graph on 2^n vertices, where the weights are complex numbers and the weight of an edge from u to v is the complex conjugate of the weight of the edge from v to u. We call the undirected graph formed by connecting two vertices if and only if the edge between them has nonzero weight the *graph of the Hamiltonian*.

A Hamiltonian is said to be d-sparse if it has at most d nonzero entries in each row (i.e., the maximum degree of its graph is d). We often associate properties of the graph of a Hamiltonian with the Hamiltonian itself. For instance, we might say "H is a forest," meaning that the graph of H is a forest.

A *star graph* is a tree in which one vertex (called the center) is connected to all the other vertices and there are no other edges. In other words, it is a complete bipartite graph $K_{1,r}$. We call a forest in which each tree is a star graph a *galaxy*.

A directed graph is a *directed forest* (*directed tree*) if its undirected graph is a forest (tree). A directed tree is an *arborescence* if it has a unique root v such that all edges point away from v. Alternately, there is exactly one directed path from v to any other vertex u. In an arborescence, the edges are always directed from the parent to the child. A directed forest in which each tree is an arborescence is called a forest of arborescences.

We use several matrix norms in our analysis. These include the spectral norm, $\|H\| := \max_{\|v\|=1} \|Hv\|$; the maximum entry norm, $\max(H) := \max_{ij} |H_{ij}|$; and the maximum column norm, $\mathrm{mcn}(H) := \max_j \|He_j\|$, where e_j is the jth column of the identity matrix.

3   Problem Description and Previous Results

The problem is to approximately implement the unitary e^{-iHt} for a d-sparse and efficiently row-computable N-dimensional Hamiltonian H for time t. As input, we are given black-box access to H, and the values of d, t, and N. Since the Hamiltonian is sparse and efficiently row-computable, there is a convenient black-box formulation of the problem that abstracts away the details of computing matrix entries and locations. The Hamiltonian is provided as a black-box function f, which accepts a row index and an integer $i \in \{1, 2, \ldots, d\}$ and outputs the column index and matrix element corresponding to the ith nonzero entry in that row, if one exists. More precisely, if the nonzero elements in row x are $y_1, y_2, \ldots, y_{d_x}$, where $d_x \leq d$ is the degree of x, then $f(x, i) = (y_i, H_{x,y_i})$ for $i \leq d_x$ and $f(x, i) = (x, 0)$ for $i > d_x$. This black box can be implemented efficiently if the Hamiltonian to be simulated is sparse and efficiently row-computable.

For each row x, we allow the order in which the y_i are given by the oracle to be arbitrary (but fixed). We do not assume that there is a convenient ordering, such as the increasing order of labels. To use the black box in a quantum circuit, we define an equivalent unitary matrix U_f which performs the operation $U_f |x, i, 0\rangle = |x, i, f(x, i)\rangle$.

Let us denote the minimum number of queries to U_f required to approximately simulate e^{-iHt} (up to error ϵ, as quantified by the trace distance) by $Q(H, t)$. A common approach to this problem breaks it into two subproblems, which we call the Hamiltonian decomposition problem and the Hamiltonian recombination problem. First the Hamiltonian is decomposed into a sum of easy-to-simulate Hamiltonians; then these Hamiltonians are simulated for short times in a specific manner so that the overall simulation is approximately the same as that of H.

Since we will also follow the decomposition–recombination strategy, we review this approach as applied in Ref. [3]. The given Hamiltonian H is decomposed into a sum of m Hamiltonians, $H = \sum_{j=1}^{m} H_j$. Let $Q(H_j)$ denote the number of queries required to simulate H_j for time t' given black-box access to H. In general, the number of queries required might depend on t', but in the simulations used here $Q(H_j)$ is independent of t'. Note that $Q(H_j)$ includes the number of queries required to decompose H into H_j as well as to simulate H_j. In Ref. [3], the Hamiltonians H_j are 1-sparse, and their decomposition uses $O(\log^* N)$ queries[1] to a black box for H. Since a 1-sparse Hamiltonian can be simulated with 2 queries given an oracle for the 1-sparse Hamiltonian [7,9], $Q(H_j) = O(\log^* N)$. More precisely,

---

[1] The function $\log^*$ is defined by $\log^* N = 0$ if $N \leq 1$ and $\log^* N = 1 + \log^* \log N$ if $N > 1$.

Theorem 1 (Hamiltonian edge decomposition [3]). *If H is an $N \times N$ Hamiltonian with maximum degree d, then there exists a decomposition $H = \sum_{j=1}^{m} H_j$, where each H_j is 1-sparse, such that $m = 6d^2$ and each query to any H_j can be simulated by making $Q(H_j) = O(\log^* N)$ queries to H.*

These Hamiltonians are then recombined using the Lie–Trotter formula, which expresses the time evolution due to H as a product of time evolutions due to the H_j. The unitary e^{-iHt} is approximated by a product of exponentials $e^{-iH_j t'}$, such that the maximum error in the final state does not exceed ϵ. We want to upper bound the number of exponentials required, $N_{\exp}$. Reference [3] proves the following.

Theorem 2 (Hamiltonian recombination [3]). *Let k be any positive integer. If $H = \sum_{j=1}^{m} H_j$ is a Hamiltonian to be simulated for time t by a product of exponentials $e^{-iH_j t'}$, and the permissible error (in terms of trace distance) is bounded by $\epsilon \leq 1 \leq 2m5^{k-1} \|H\| t$, then the number of exponentials required, $N_{\exp}$, is bounded by*

$$N_{\exp} \leq 5^{2k} m^2 \|H\| t \left(\frac{m \|H\| t}{\epsilon} \right)^{1/2k}. \tag{1}$$

Using the upper bound on the number of exponentials and the number of queries needed to simulate any exponential, the total number of queries needed to simulate the Hamiltonian H satisfies $Q(H,t) \leq N_{\exp} \times \max_j Q(H_j)$. With $Q(H_j) = O(\log^* N)$ and $m = 6d^2$, we get

$$Q(H,t) = O\left(5^{2k} d^4 (\log^* N) \|H\| t \left(\frac{d^2 \|H\| t}{\epsilon} \right)^{1/2k} \right). \tag{2}$$

We see that $Q(H,t)$ is almost linear in t, which is almost optimal due to a no–fast–forwarding theorem [3]. However, the dependence on d is not optimal. In the present paper we improve the dependence on d without affecting the other terms. The dependence on d has been improved in other approaches, but only at the expense of a worse dependence on the error ϵ [10,11].

We propose a new algorithm for solving the Hamiltonian decomposition problem. This strategy breaks up the Hamiltonian into only $m = 6d$ parts, but increases $Q(H_j)$ to $O(d + \log^* N)$, improving the overall dependence on d and N.

4   Hamiltonian Decomposition

The Hamiltonian decomposition problem is the problem of decomposing a Hamiltonian H into a sum of m Hamiltonians H_j such that given a label $1 \leq j \leq m$ and a time t', the unitary $e^{-iH_j t'}$ can be efficiently simulated.

We solve this problem by decomposing the Hamiltonian into $m = 6d$ galaxies. To achieve this, we first decompose the given graph into d forests using the forest decomposition technique of Paneconesi and Rizzi [12]. The idea is to assign one of at most d colors to each edge of the graph (not necessarily a proper edge coloring) such that the edges of any particular color form a forest. Not only is this decomposition possible, but it has some special properties that are required later in Lemma 2.

Lemma 1 (Forest decomposition). *For any Hamiltonian H of maximum degree d, there exists a decomposition $H = \sum_{c=1}^{d} H_c$ and an assignment of directions to the edges such that each H_c is a forest of arborescences. Furthermore, given a color c and a vertex v, we can determine v's parent in H_c with one query (or determine that it is a root) and with $O(d)$ queries we can determine the list of edges in H_c incident on v.*

Proof. We first describe a procedure that assigns a color c to each edge. H_c then consists of all edges colored c. To color the edges, every vertex proposes a color for each edge incident on it using the oracle in the following way: if $f(x, i) = (y, H_{x,y})$, then x proposes color i for the edge xy. Similarly, y proposes a color for the edge xy. The edge is now colored using the proposal of the vertex with higher label (i.e., if $x > y$ then the edge xy is colored with x's proposal). This coloring uses d colors, which is optimal up to constants since a d-sparse graph can have $dn/2$ edges, but forests have at most $n - 1$ edges.

Now we assign directions to the edges and show that each H_c has no cycles, which shows that each H_c is a directed forest. The edge xy is directed from x to y if $x < y$. This choice of directions results in a directed acyclic graph, which has no directed cycles. To rule out non-directed cycles, we note that any such cycle must contain a vertex v for which both the edges of the cycle point toward v. This means the label of v is greater than that of its two neighbors. Thus the color of these edges was decided by v, which cannot happen since vertices propose different colors for different edges.

To show that each tree in H_c is an arborescence, we show that it has a unique root. Observe that a directed tree with more than one root must have a vertex with more than one parent. This again leads to the situation where a vertex has two incoming edges of the same color, which is not possible since these edges are colored by this vertex's proposal.

To show that the parent of a vertex can be determined with one query, note that if p_v is the parent of vertex v in H_c, then the edge from p_v to v must be directed toward v. Thus the color of this edge is decided by v. If this edge is in H_c, it is colored c. So if v has a parent, it must be the cth neighbor of v. With one query to the oracle, we can determine the cth neighbor of v. If there is no such neighbor, this vertex has no parent and is a root in H_c. Otherwise the output contains the label of the parent.

Finally, we show how to determine the list of edges in H_c incident on x with $O(d)$ queries. First we query the oracle at most d times to get the labels of all the neighbors of x. For a neighbor y where $y < x$, the edge between x and y

is colored by c only if y is x's parent in H_c. Thus we can discard all edges xy where $y < x$ but y is not the parent of x. When $y > x$, an edge between x and y is colored c only if x is y's parent in H_c, and it takes one query to verify this for each y. Thus with at most d additional queries we can determine if all such edges are colored c. □

This lemma shows how to decompose a Hamiltonian into directed forests. Let T be the Hamiltonian of such a forest. We will decompose T into a sum of 6 galaxies, $T = T_1 + T_2 + \cdots + T_6$. This is achieved by using an extension of the "deterministic coin tossing" protocol of Cole and Vishkin [13] by Goldberg, Plotkin and Shannon [14]. Their protocol gives a proper vertex coloring of an arborescence using only 6 colors making $O(\log^* N)$ queries. Vertex coloring a directed forest of arborescences gives a galaxy decomposition of the forest, since all the edges that point to vertices of a particular color form a galaxy.

Lemma 2 (Vertex coloring a forest). *If T is a forest of arborescences, and the parent of a vertex can be determined with one query to an oracle for T, then there exists a proper vertex coloring of T using 6 colors, such that the color of any vertex can be determined by making $O(\log^* N)$ queries.*

Proof. We first describe the vertex-coloring procedure for the forest. A simple observation is that we already possess a vertex coloring of the forest: the labels of the vertices. This is a trivial proper vertex coloring using N colors. Now we use a procedure that decreases the number of colors used by a logarithmic factor. Then we can run several rounds of this procedure to decrease the number of colors down to 6. Let $c_j(x)$ be the color assigned to vertex x at the beginning of the jth round of the procedure. At the beginning of the first round, we have $c_1(x) = x$.

Let x be a vertex with parent p_x. Assume that we started with a proper vertex coloring at the beginning of round j. Since we have a proper coloring, $c_j(x) \neq c_j(p_x)$. Let k be the index of the first bit at which x and p_x differ, and let b be the value of the kth bit of x. The new color for vertex x is the concatenation of k and b, denoted (k, b). If x is the root, we take $k = 0$. We claim that if each vertex performs this procedure, the result is a proper vertex coloring.

For a contradiction, suppose there are two adjacent vertices that have been assigned the same color in round j. Without loss of generality, one of them is the parent of the other, so let them be y and its parent p_y. Since we started with a proper coloring at the beginning of round j, $c_j(y) \neq c_j(p_y)$, but now $c_{j+1}(y) = c_{j+1}(p_y)$. Let $c_{j+1}(y) = (k, b)$ where by definition k is the bit at which $c_j(y)$ and $c_j(p_y)$ differ, and b is the value of the kth bit of $c_j(y)$. Since $c_{j+1}(p_y)$ also equals (k, b), the kth bit of $c_j(p_y)$ is b. But $c_j(y)$ and $c_j(p_y)$ are supposed to differ at the kth bit, so this is a contradiction. Therefore the coloring procedure is valid.

It remains to show that if the colors of the vertices are updated in this way, we reduce the number of colors to 6 in $O(\log^* N)$ rounds. If L_j is the number of bits used to represent colors at the beginning of round j, then $L_{j+1} = \lceil \log(L_j) \rceil + 1$.

Initially, $L_1 = \lceil \log N \rceil$. This recurrence relation can be solved to yield $L_j \leq 3$ when $j = \log^* N$ [14]. Further rounds cannot decrease L_j below 3, since $L_{j+1} = L_j$ when $L_j = 3$. A length of 3 bits allows the use of only 8 colors. Now we run the procedure once more. Since there are 3 possible values for k, and 2 for b, there are at most 6 different colors. The total number of rounds is now $\log^* N + 1$.

To show that the color of a vertex can be determined with $O(\log^* N)$ queries, we note that the color of vertex x at the end of the first round depends solely on x and p_x. In general, the color of vertex x at the end of j rounds depends only on its first j ancestors. To determine x's color after $\log^* N + 1$ rounds, we need the labels of its $\log^* N + 1$ ancestors, which can be found with $\log^* N + 1$ queries, since the parent of a vertex can be found with one query.   □

We have shown that a Hamiltonian can be decomposed into d forests of arborescences, each of which can be vertex-colored with 6 colors. If we consider all the edges of one of the d forests that point to a vertex of a particular color, this graph is a galaxy. So this decomposes the original Hamiltonian into $6d$ galaxies. For this particular decomposition of the Hamiltonian to be useful, we need to show that galaxies can be simulated easily.

Theorem 3 (Galaxy simulation). *If H_j is a Hamiltonian whose graph is a galaxy of maximum degree d, and the oracle can identify which vertices are centers of stars, then the unitary operator $e^{-iH_j t}$ can be simulated using $O(d)$ calls to an oracle for H_j.*

Proof. The key idea is that given a vertex v, we can learn everything about the star to which v belongs in $O(d)$ queries. If v is the center of the star, the oracle identifies it as the center, so we can query all its neighbors to learn everything about the star with at most d queries. If v is not the center, we can determine the center, which is the only neighbor of v, with only one query, and then learn the rest of the star with at most d queries.

Let $R(x)$ denote all the information about the star to which x belongs: the label of the center, the labels of the other vertices in some fixed order, and the weights of all the edges. It is essential that $R(x)$ depend only the star and not the particular vertex x chosen from the star, so that if x and y belong to the same star then $R(x) = R(y)$. Since we know that $R(x)$ can be computed with $O(d)$ queries, we can implement the unitary U given by $U |x, 0\rangle = |x, R(x)\rangle$ with $O(d)$ queries.

The Hamiltonian we are trying to simulate, H_j, is a galaxy. Thus, if c is the center of a star, and its neighbors are y_i with edge weights w_i, then $H_j |c\rangle = \sum_i w_i |y_i\rangle$. If x is not the center of a star, and the edge between x and the center c has weight w_x, then $H_j |x\rangle = w_x |c\rangle$. Let K be a Hamiltonian which is similar to H_j, but acts on the input state $|x, R(x)\rangle$ instead of $|x\rangle$. That is, $K |c, R(c)\rangle = \sum_i w_i |y_i, R(y_i)\rangle$ when c is the center, and $K |x, R(x)\rangle = w_x |c, R(c)\rangle$ otherwise. Note that although the second register looks different, it is unaffected by K since $R(x)$ depends only on the star and not the vertex. Combining K with the unitary U above, we see that $H_j = U^\dagger K U$. In words, U first computes all the information about the star in another register, K performs the required Hamiltonian, and the $U^\dagger$ uncomputes the second register, which was unaffected by K.

This simulation is efficient since K can be simulated efficiently. More importantly for our purposes, K requires no queries to implement, since all the information about the star is already present in the second register. Thus the operation $H_j = U^\dagger K U$ requires only as many queries as U and $U^\dagger$ require, which is $O(d)$.   □

Combining Lemma 1, Lemma 2, and Theorem 3 gives our Hamiltonian decomposition theorem.

Theorem 4 (Hamiltonian star decomposition). *There exists a decomposition $H = \sum_{j=1}^{m} H_j$, where each H_j is a galaxy, such that $m = 6d$ and each galaxy H_j can be simulated for time t' using $Q(H_j) = O(d + \log^* N)$ queries to an oracle for H.*

Proof. From Lemma 1, Lemma 2, and Theorem 3, we know that the claimed decomposition is possible. It remains to show that any H_j can be simulated for time t' using $O(d + \log^* N)$ queries.

To show this, let us implement H_j on the basis state $|x\rangle$. If the implementation is correct on all basis states, it is correct for all input states by linearity. We are given $1 \le c \le d$ and $1 \le t \le 6$, which together form the index j. We want to simulate the galaxy formed by edges in H_c directed toward vertices colored t by the vertex coloring algorithm of Lemma 2.

From the proof of Theorem 3, it is clear that if we can compute $R(x)$, then we can implement U, and thereby simulate the desired Hamiltonian. $R(x)$ contains all the information about the star to which x belongs. Using the result of Lemma 1, we can determine the list of x's neighbors in H_c using $O(d)$ queries. By the result of Lemma 2, with $O(\log^* N)$ queries we can determine x's color according to the vertex coloring algorithm.

If x's color is not t, then x must be the center of a (possibly empty) star in H_j. The only edges in this star point toward vertices of color t, so we compute the colors of all the children of x in H_c. These can be computed using only the labels of their $\log^* N + 1$ nearest ancestors, which are all common ancestors. Thus we can compute the colors of all of x's children using $O(\log^* N)$ queries in total. Now we know the star around x, and thus $R(x)$, using $O(d + \log^* N)$ queries.

If x's color is t, then x's parent is the center of star. The parent of x, p_x, can be determined with one query. Since x and p_x are in the same star, $R(x) = R(p_x)$. Since p_x is the center of a star, we can compute $R(p_x)$ as described above; thus we can also compute $R(x)$.

We have shown that for any x, we can compute $R(x)$ with $O(d + \log^* N)$ queries. Thus the unitary U in the proof of Theorem 3 can be simulated with $O(d + \log^* N)$ queries. By Theorem 3, this means we can implement H_j with $O(d + \log^* N)$ queries, as claimed.   □

Now we can use our Hamiltonian decomposition theorem with the Hamiltonian recombination theorem (Theorem 2). Since we have $Q(H_j) = O(d + \log^* N)$ from

Theorem 4 and $m = 6d$ from Lemma 1 and Lemma 2, we get our final result using $Q(H, t) \leq N_{\exp} \times \max_j Q(H_j)$:

$$Q(H, t) = O\left(5^{2k} d^2 (d + \log^* N) \|H\| t \left(\frac{d \|H\| t}{\epsilon}\right)^{1/2k}\right). \tag{3}$$

When compared with the query complexity of (2), we see that this improves the scaling with d. Furthermore, when $d = \Omega(\log^* N)$, which is likely to be the case when d is not constant, (3) has no $\log^* N$ term: the scaling (in terms of d and N) is $(d^3)^{1+o(1)}$, as compared to (2) which scales like $(d^4 \log^* N)^{1+o(1)}$.

5   Remarks and Conclusion

So far, we have measured the size of H using the spectral norm $\|H\|$. However, if we express the simulation complexity in terms of a different norm, then both Theorem 2 and equation (3) can be improved to give slightly better bounds.

In the proof of Theorem 2, $\|H\|$ is used as a simple upper bound for $\max_j \|H_j\|$. However, omitting this step gives a slightly stronger version of Theorem 2 with $\|H\|$ replaced by $\max_j \|H_j\|$. For a 1-sparse Hamiltonian, $\|H_j\| = \max(H_j) \leq \max(H)$ [15], so $\|H\|$ can be replaced by $\max(H)$ in (2). However, this also leads to an improvement of (3). When H_j is a galaxy, $\|H_j\| = \text{mcn}(H_j)$ [15], and since H_j is entry-wise upper bounded by H, $\text{mcn}(H_j) \leq \text{mcn}(H)$. Thus $\|H\|$ can be replaced with $\text{mcn}(H)$ in (3). To directly compare the two simulations, we can apply the bound $\text{mcn}(H) \leq \sqrt{d} \max(H)$ [15] to express both query complexities in terms of $\max(H)$. In these terms, we still find that star decomposition improves over edge coloring: our algorithm uses at most $(d^{2.5}(d + \log^* N) \max(Ht))^{1+o(1)}$ queries, whereas the algorithm of Ref. [3] scales like $(d^4 (\log^* N) \max(Ht))^{1+o(1)}$.

In conclusion, we have described a Hamiltonian decomposition technique that reduces the query complexity of simulating sparse Hamiltonians. By the degree-dependent lower bounds established in Ref. [15], we know that query complexities scaling like $\|H\|$ or $\sqrt{d} \max(H)$ cannot be achieved. It would be interesting to see if the Hamiltonian decomposition–recombination framework can be used to further reduce the dependence on d while keeping a similar dependence on the error ϵ, or to establish stronger limitations on the simulation of sparse Hamiltonians taking error dependence into account.

References

1. Lloyd, S.: Universal quantum simulators. Science 273(5278), 1073–1078 (1996)
2. Aharonov, D., Ta-Shma, A.: Adiabatic quantum state generation and statistical zero knowledge. In: Proc. 35th STOC, pp. 20–29. ACM, New York (2003)
3. Berry, D., Ahokas, G., Cleve, R., Sanders, B.: Efficient quantum algorithms for simulating sparse Hamiltonians. Commun. Math. Phys. 270(2), 359–371 (2007)
4. Feynman, R.: Simulating physics with computers. Int. J. Theor. Phys. 21(6), 467–488 (1982)

5. Farhi, E., Goldstone, J., Gutmann, S., Sipser, M.: Quantum computation by adiabatic evolution (2000), ArXiv preprint quant-ph/0001106
6. Farhi, E., Gutmann, S.: Analog analogue of a digital quantum computation. Phys. Rev. A 57(4), 2403–2406 (1998)
7. Childs, A.M., Cleve, R., Deotto, E., Farhi, E., Gutmann, S., Spielman, D.A.: Exponential algorithmic speedup by a quantum walk. In: Proc. 35th STOC, pp. 59–68. ACM, New York (2003)
8. Farhi, E., Goldstone, J., Gutmann, S.: A Quantum Algorithm for the Hamiltonian NAND Tree. Theory of Computing 4, 169–190 (2008)
9. Childs, A.M.: Quantum information processing in continuous time. PhD thesis, Massachusetts Institute of Technology (2004)
10. Childs, A.M.: On the Relationship Between Continuous- and Discrete-Time Quantum Walk. Commun. Math. Phys. 294(2), 581–603 (2010)
11. Berry, D.W., Childs, A.M.: The quantum query complexity of implementing black-box unitary transformations (2009), ArXiv preprint arXiv:0910.4157
12. Panconesi, A., Rizzi, R.: Some simple distributed algorithms for sparse networks. Distrib. Comput. 14(2), 97–100 (2001)
13. Cole, R., Vishkin, U.: Deterministic coin tossing with applications to optimal parallel list ranking. Inf. Control 70(1), 32–53 (1986)
14. Goldberg, A.V., Plotkin, S.A., Shannon, G.E.: Parallel symmetry-breaking in sparse graphs. SIAM J. Discrete Math. 1(4), 434–446 (1988)
15. Childs, A., Kothari, R.: Limitations on the simulation of non-sparse Hamiltonians. Quantum Information and Computation 10, 669–684 (2010)

The Polynomial Degree of Recursive Fourier Sampling

Benjamin Johnson

School of Information,
University of California, Berkeley

Abstract. We present matching upper and lower bounds for the "weak" polynomial degree of the recursive Fourier sampling problem from quantum complexity theory. The degree bound is $h + 1$, where h is the order of recursion in the problem's definition, and this bound is exponentially lower than the bound implied by the existence of a BQP algorithm for the problem. For the upper bound we exhibit a degree-$h + 1$ real polynomial that represents the problem on its entire domain. For the lower bound, we show that any non-zero polynomial agreeing with the problem, even on just its zero-inputs, must have degree at least $h + 1$. The lower bound applies to representing polynomials over any Field.

1 Introduction

The recursive Fourier sampling problem holds a central place in the folklore and history of quantum complexity theory. First defined in the very same 1993 preliminary abstract [1] of "Quantum Complexity Theory" [2] that defined BQP and EQP, and proved the existence of a universal quantum Turing machine, the problem was used to sketch the first oracle separation of BQP from BPP. This result was among the first in a long line of later results relating various quantum complexity classes to their classical counterparts. (e.g. [3][4][5]).

The continued importance of the problem goes beyond its historical significance, as several conjectures regarding the problem's classical complexity class placement remain open. It has been conjectured that the recursive Fourier sampling problem is not in the class AM, or even the polynomial-time hierarchy, PH [1][6]. Yet not only do these specific conjectures remain open, but the more general conjecture of the existence of any oracle placing BQP outside AM or PH remain open [7].

An affirmation of these longstanding conjectures would follow directly from an appropriate constant depth circuit size lower bound on the recursive Fourier sampling problem. Unfortunately, as is well-known, the usual circuit lower bound techniques based on polynomial degree require a problem's degree to be higher than it actually is for any problem in BQP [8][4]. The results in this paper show further that the polynomial degree of the recursive Fourier sampling problem is exponentially lower than many other problems in BQP, illustrating how thoroughly those standard circuit lower bound techniques fail to work on this

W. van Dam et al. (Eds.): TQC 2010, LNCS 6519, pp. 104–112, 2011.
© Springer-Verlag Berlin Heidelberg 2011

problem. Our main result turns this lemon into lemonade, exploiting the low
degree of the problem to offer a non-trivial lower bound on the degree itself.
The result may also be of independent interest, as it introduces a lower bound
technique for recursively defined problems.

We begin in section 3 by defining the recursive Fourier sampling problem and
introducing relevant notation. We review preliminary material about polynomials and define our notion of polynomial representation in section 2. Section 4
outlines the construction of the problem's representing polynomial. In section 5
we state and prove our central result – the polynomial degree lower bound. We
conclude in section 6.

2   Preliminaries

In this preliminary section we include not only the standard notation for polynomials representing boolean functions, but also a few new definitions relating
to representation on reduced domains.

Let $\mathbb{K}$ be any Field. Any N-variate polynomial P over $\mathbb{K}$ can be written in
canonical form as $P(x_1, \ldots, x_N) = \sum_{S \subseteq \{1, \ldots, N\}} \alpha_S \prod_{i \in S} x_i$ where $\alpha_S \in \mathbb{K}$ for
each S.

By a **monomial** M, we simply mean a product of variables (of the form
$\prod_{i \in S} x_i$ for some $S \subset \{1, \ldots, N\}$).

We will say that such a monomial $M = \prod_{i \in S} x_i$ occurs in P if there is
$T \subseteq \{1, \ldots, N\}$ such that $S \subseteq T$ and α_T (the coefficient of $\prod_{i \in T} x_i$ in the
canonical representation of P) is non-zero. Note that "M occurs in P" is a
weaker condition than "the coefficient of M in P is non-zero." To say that "M
occurs in P" just means that M divides some (possibly larger) monomial whose
coefficient in P is non-zero.

For a monomial M, and polynomial P, we define the total coefficient of M
in P to be the unique multi-linear polynomial Q defined by the conditions $P =$
$MQ + R$, where Q, R are multi-linear polynomials, none of the x_i with $i \in S$
occur in Q, and M does not occur in R. Alternatively, the total coefficient of M
in P is the quotient obtained when we divide M into P using polynomial long
division.

We define a strict partial order $\prec$ on the set of all polynomials on N variables
over $\mathbb{K}$ by $P \prec P'$ if and only if P can be obtained from P' by substituting
the constants 0 or 1 for some of the variables in some of the monomials of P'.
For example, $x_0 x_1 x_2 + x_3 \prec x_0 x_1 x_2 + x_2 x_3$. Notice that $\prec$ is well-founded, since
replacing a variable with a constant in any monomial decreases the sum, over
all monomials M occurring in P, of the number of variables in M.

Let $D \subseteq \{0, 1\}^N$. We say that P is reduced with respect to D, if there does
not exist a polynomial P' that both agrees with P on D, and satisfies $P' \prec P$.

For $X \in \{0, 1\}^N$, and for $I \subseteq \{0, \ldots, N - 1\}$, let X^I denote the element
of $\{0, 1\}^N$ obtained from X by flipping the bits in I. If $S = \{x_i : i \in I\}$ is
a collection of boolean variables indexed by I, we will sometimes write X^S to
mean the same thing as X^I.

Let F be any partial boolean function $F : \{0,1\}^N \rightarrow \{0,1,*\}$; and let P be an N-variate multilinear polynomial over a Field $\mathbb{K}$. For $b \in \{0,1\}$, we say that P *weakly b-represents* F *over* $\mathbb{K}$ if for every b-input of F we have $P(X) = b$, and for at least one $(1\text{-}b)$-input of F, we have $P(X) \neq b$.

3   Definition of Recursive Fourier Sampling

For each pair of non-negative integers n and h, let $N_{n,h}$ be the recursively-defined function satisfying the initial condition $N_{n,0} = 1$ and the recurrence relation $N_{n,k+1} = 2^n \cdot N_{n,k} + 2^n$. Then for each pair of non-negative integers n and h, we define the partial boolean function:

$$RFS_{n,h} : \{0,1\}^{N_{n,h}} \rightarrow \{0,1,*\} \tag{1}$$

as follows. The definition is by recursion on h.

Let $RFS_{n,0}$ be the identity function on $\{0,1\}$. Now assume that $RFS_{n,k}$ has been defined. To define $RFS_{n,k+1}$, we construe the $N_{n,k+1}$-bit input as a pair $(\langle G_\sigma \rangle_{\sigma \in \{0,1\}^n}, g)$, where for each $\sigma \in \{0,1\}^n$, G_σ is an input to $RFS_{n,k}$ of length $N_{n,k}$; and $g : \{0,1\}^n \rightarrow \{0,1\}$. Then define

$$RFS_{n,k+1}(\langle G_\sigma \rangle_{\sigma \in \{0,1\}^n}, g) = \begin{cases} g(s) & \text{if } \forall \sigma \in \{0,1\}^n \ RFS_{n,k}(G_\sigma) = \sigma \cdot s \\ * & \text{otherwise} \end{cases} \tag{2}$$

Define $DRFS_{n,h}$ to be $RFS_{n,h}^{-1}[\{0,1\}]$. We may also sometimes refer to $DRFS_{n,h}$ as the domain of $RFS_{n,h}$.

Recursive Fourier Sampling Trees. We may visualize $RFS_{n,h}$ in terms of a 2^n-branching tree $T_{n,h}$ of height h, and we will use the tree to develop some notation to use in proof.

We picture the tree with the root at the top at height h, and the leaves at the bottom at height 0. A convenient way to label the vertices in such a tree is with sequences of boolean variables, the length of which corresponds to the distance from the root vertex. The root vertex will have an empty label, the children of the root vertex will have a label x_h of length n, and on down the tree, until each leaf vertex has a label $x_1, \ldots, x_h$ of length hn. If σ is a vertex at height k, it has a label of length $(h-k)n$, which we will reference in n-bit blocks as $x_{k+1}, \ldots, x_h$. Suppose that σ at height k has label $a_{k+1}, \ldots, a_h$. Then if $k < h$ we can reference its unique parent as $a_{k+2}, \ldots, a_h$; and if $k > 0$ we can reference each of its 2^n children with strings of the form $x_k, a_{k+1} \ldots, a_h$.

Relating this to $RFS_{n,h}$, we associate a boolean function g_σ and a secret string s_σ to each vertex σ. If ρ is a leaf vertex at height 0, we associate a single bit, which we call g_ρ. [For notational convenience, we may construe g_ρ as a function on $\{0,1\}^0$, and may refer to its empty input variable as z_0]. If σ is a non-leaf vertex at height k, we associate both an n-bit function, which we call g_σ, and an n-bit string s_σ. We reference the input to g_σ with a variable y_k of length n.

We may think of the task of computing $RFS_{n,h}$ as "solving this tree" under the following interpretation: For each non-leaf vertex $\sigma = a_{k+1}, \ldots, a_h$ at height k, there is both a challenge and a promise. The challenge at σ is to determine $g_\sigma(s_\sigma)$, while the promise for σ is that for every $y_k \in \{0,1\}^n$, we have $g_{y_k,\sigma}(s_{y_k,\sigma}) = y_k \cdot s_\sigma$. To "solve the tree" we must answer the challenge for the root vertex τ, by accessing only a limited amount of the functions g_σ, typically by querying some number of bits that is a polynomial function of n.

4   Upper Bound

Theorem 1. *There is a real polynomial of degree $h+1$ that represents the function $RFS_{n,h}$ exactly on its domain.*

Proof. Sketch. It is easily verified by inspection that the real polynomial

$$Q(f,g) = \frac{1}{2^n} \sum_{y \in \{0,1\}^n} \sum_{x \in \{0,1\}^n} (-1)^{x \cdot y}(1 - 2f(x))g(y) \tag{3}$$

represents the Fourier sampling problem $(RFS_{n,1}(f,g))$ exactly on its domain. When we apply to Q the recursive definition for $RFS_{n,h}$, we obtain a degree $h+1$ real polynomial that represents $RFS_{n,h}$ on its domain. See [9] for a more thorough construction along with examples for small values of h.

5   Lower Bound

5.1   Main Result and Proof Outline

The following theorem is our main result of the paper. We state the theorem here, and outline the proof strategy. The details are carried out in the two lemmas that follow, and a wrap-up at the end of the section. [Recall that P weakly 0-represents F over $\mathbb{K}$ if for every 0-input of F we have $P(X) = 0$, and for at least one 1-input of F, we have $P(X) \neq 0$.]

Theorem 2. *(Main Result) Let $\mathbb{K}$ be a Field. If $h < 2^n$ and P is any N-variate multilinear polynomial that weakly 0-represents (or 1-represents) $RFS_{n,h}$ over $\mathbb{K}$, then P has degree at least $h+1$.*

Sketch of the proof: Assume without loss of generality that P is reduced with respect to $DRFS_{n,h}$. If P weakly 0-represents (or 1-represents) $RFS_{n,h}$, it must contain at least one variable of the form $g_\tau(a)$ where τ is the root vertex at height h. Let M be a monomial in P of maximal length that contains only variables of this form, say $M = \prod_{j=1}^{k} g_\tau(a_j)$ for some $a_1, \ldots, a_k \in \{0,1\}^n$. If $k \geq h+1$ we are done, so assume $1 \leq k \leq h < 2^n$, and consider the quotient Q defined by $P = QM + R$. Since P is reduced, Q cannot be zero on all of $DRFS_{n,h}$, and Q does not contain any variables of the form $g_\tau(a)$. We use this to show that Q weakly 0-represents the partial function on $DRFS_{n,h}$ defined by

$$F(G) = \begin{cases} 0 \text{ if } \bigwedge_{j=1}^{k} s_\tau^G \neq a_j \\ 1 \text{ if } \bigvee_{j=1}^{k} s_\tau^G = a_j \end{cases}$$. (Note that this function is always well-defined

and non-constant because $k < 2^n$). Repeating the procedure for Q, we obtain a non-zero polynomial Q_1, which contains no variables of the form $g_\sigma(b)$ with σ any vertex at height $h - 1$ or h, and that weakly-represents a function on $DRFS_{n,h}$ whose value depends on s_σ, with σ a vertex at height $h - 1$. An inductive argument establishes that this procedure can be carried out h times, while after each step another monomial of degree at least one has been factored out. We conclude from this that the original polynomial P must have degree at least $h + 1$.

5.2   Reduction and Invariance Lemmas

We require two key lemmas that will aid in proving our polynomial degree lower bound for $RFS_{n,h}$. Before summarizing these results, we remind the reader what it means for a polynomial to be reduced relative to a given domain. Loosely speaking, a polynomial P is reduced relative to a domain $D \subseteq \{0,1\}^N$ if every polynomial P' that can be obtained from P by replacing variables with constants in a given sub-polynomial of P, satisfies $P'|_D \neq P|_D$. The formal (longer but clearer) definition is given in section 2.

The first lemma (which we may refer to as the "Reduction Lemma") says that if a multilinear polynomial P is non-zero and reduced on a domain D, then the total coefficient of each monomial of P is also non-zero and reduced on D. Without the condition of being reduced, the non-zero part of the lemma would be false, because a non-reduced polynomial P that is non-zero on D can have sub-polynomials that are zero on D. The reduction lemma tells us that if we start with a *reduced* polynomial, and we notice that a monomial M occurs in P, then we can be sure that the total coefficient of M in P is doing something non-zero on D.

The second lemma (which we may refer to as the "Invariance Lemma") says that for any fixed input $G \in \{0,1\}^N$ and monomial $M = \prod_{i=1}^{k} x_i$, if the value of $P(G)$ is invariant under bit changes to G for any combination of the k bits mentioned in M, then the total coefficient Q of M in P evaluates to zero on input G. This lemma allows us to use an invariance property of P on a collection of inputs, to deduce a 0-invariance property on an appropriately-chosen sub-polynomial of P on the same collection of inputs. When combined with the reduction lemma, this allows us to factor monomials in a way that keeps two properties: *zero on a sufficiently large set of inputs*, and *non-zero on at least one input*.

These two lemmas give us a way to maintain a weak representation condition through the process of factoring monomials.

Lemma 1. *(Reduction Lemma) Let P be a non-zero N-variate multi-linear polynomial over $\mathbb{K}$ that is reduced with respect to $D \subseteq \{0,1\}^N$. Let $M = \prod_{i \in I} x_i$ be a monomial that occurs in P, and let Q be its total coefficient. Then*

1. Q is reduced with respect to D.
2. There is an $X \in D$ such that $Q(X) \neq 0$. [Note: for Q reduced, this is equivalent to saying Q is non-zero]

Proof.   1. If Q is not reduced, then there is a polynomial Q' agreeing with Q on D such that $Q' \prec Q$. Suppose $P = MQ + R$. Then letting $P' = MQ' + R$, we have P' agreeing with P on D, and $P' \prec P$. This contradicts our assumption that P was reduced on D.

2. If $Q(X) = 0$ for every $X \in D$, then the polynomial R described above agrees with P on D and (since Q was assumed non-zero) satisfies $R \prec P$. This contradicts our assumption that P was reduced on D.

Lemma 2. *(Invariance Lemma) Let P be a real multilinear polynomial in N variables $x_1, \ldots, x_N$, and let M be a monomial of length k, say $M = \prod_{j=1}^{k} x_j$. Let $D \subseteq \{0,1\}^N$, and let $G \in D$ be such that $P(G)$ remains constant when we independently toggle any combination of the bits $x_1, \ldots x_k$. [The notation we will use for this is that $P(G) = P(G^S)$ for every subset $S \subseteq \{x_1, \ldots, x_k\}$]. Let us factor P as $P = MQ + R$ where none of the $x_1, \ldots, x_k$ occur in Q, and M does not occur in R. [Note that Q and R are uniquely determined by these conditions]. Then $Q(G) = 0$.*

Proof. Fix all the variables in P according to G, except for $x_1, \ldots, x_k$. We obtain a new multilinear polynomial P_G in just the variables $x_1, \ldots, x_k$. The factorization $P = QM + R$ reduces to $P_G(x_1, \ldots, x_k) = Q_G \cdot M + R_G(x_1, \ldots, x_k)$. Now our condition $P(G) = P(G^S)$ for every $S \subseteq \{x_1, \ldots, x_k\}$ implies that $P_G(x_1, \ldots, x_k)$ is constant whenever the input $(x_1, \ldots, x_k)$ is boolean. Since P_G is multilinear, it must be a constant polynomial. In particular, the coefficient of $\prod_{i=1}^{k} x_i$ in P_G is zero. Observe that this coefficient is exactly $Q(G)$.

5.3   Main Lemma

**Lemma 3. *Main Lemma:* **Let $1 \leq k \leq h < 2^n$; and let Q be a multi-linear* real polynomial satisfying the following properties:*

1. Q is reduced and non-zero on $DRFS_{n,h}$.
2. No variable of the form $g_{\sigma_i}(x)$ occurs in Q for any vertex σ_i at height k or higher.
3. There exists a fixed set of at most k pairs (s_{σ_i}, a_i) with each σ_i a vertex at height k, such that $Q(G) = 0$ whenever $G \in DRFS_{n,h}$ and G satisfies $\bigwedge_{i=1}^{k}(s_{\sigma_i}^{G} \neq a_i)$.

Then Q has degree at least k.

Proof. The proof is by induction on k.

[Note that in general, there will always be a G satisfying k constraints from condition 3 of the form $\bigwedge_{i=1}^{k}(s_{\sigma_i} \neq a_i)$ as long as $k < 2^n$].

For $k = 1$, the single constraint from condition 3 is satisfiable. So together with condition 1, Q takes on at least two values, and is thus not constant. So Q has degree at least 1.

For the inductive step, assume that $k \geq 2$, and the lemma is true for $k - 1$. Let Q be a polynomial satisfying conditions 1, 2, and 3 for k.

First, we show that some variable of the form $g_\rho(x)$, with ρ a vertex at height $k - 1$, must occur in Q. The proof is by contradiction. Suppose that no variable of the form $g_\rho(x)$ occurs in Q with ρ a vertex at height $k - 1$; then combined with condition 2 for k, no variable of the form $g_\sigma(x)$ or $g_\rho(x)$ occurs in Q for σ at height k (or higher) or ρ at height $k - 1$. Consider the $X \in DRFS_{n,h}$ such that $Q(X) \neq 0$. Since Q does not depend on variables of the form $g_\sigma(x)$ or $g_\rho(x)$, we can change X on bits associated with these variables without changing the resulting value of Q on the changed X. We will do this in a way so that the changed X (call it X') meets the constraints from condition 3 for k, and thus must satisfy $Q(X') = 0$. For each σ_i mentioned in condition 3 for k, we choose any value $s_{\sigma_i}^{new}$ in $\{0,1\}^n$ different from all the a_i's that are paired with σ_i in condition 3 for k. First we will change X to obtain X' such that $s_{\sigma_i}^{X'} = s_{\sigma_i}^{new}$. For each child vertex $\rho = (\sigma_i, x)$ of σ_i, we change (if necessary) the old value $g_\rho(s_\rho^X)$ to $x \cdot s_{\sigma_i}^{new}$. We also change (if necessary) the value $g_{\sigma_i}(s_{\sigma_i}^{new})$ to match $g_{\sigma_i}(s_{\sigma_i}^X)$. After making these changes for each σ_i, the resulting X' satisfies all recursive Fourier sampling promises at all levels, and hence is in $DRFS_{n,h}$. X' also satisfies, by its construction, the constraints $\bigwedge_{i=1}^{k}(s_{\sigma_i} \neq a_i)$; hence by condition 3 for k, we have $Q(X') = 0$. We have $Q(X) \neq 0$; $Q(X) = Q(X')$, and $Q(X') = 0$, obtaining our desired contradiction. We conclude that some variable of the form $g_\rho(x)$, with ρ a vertex at height $k - 1$, must occur in Q.

Now let us factor out the largest monomial M in Q of the form $\prod_{j=1}^{r} g_{\rho_j}(b_j)$. Such a monomial with $r \geq 1$ exists by the above argument. If there is an M of this form with $r > k - 1$, then Q has degree at least k and we are done. So assume that $1 \leq r \leq k - 1$. Let Q_1 be the total coefficient of M in Q.

Now Q_1 satisfies condition 1 by the reduction lemma, and it satisfies condition 2 for $k - 1$ by its construction. We will show that Q_1 also satisfies condition 3 for $k - 1$.

Consider the set of (at most $k - 1$) pairs (ρ_j, b_j) taken from the monomial M, where each ρ_j is a vertex at height $k - 1$. Let $G \in D$ and suppose that G satisfies $\bigwedge_{j=1}^{k-1}(s_{\rho_j}^G \neq b_j)$.

First we change G on bits of the form $g_\rho(x)$ for suitable children ρ of $\sigma_1, \ldots \sigma_k$, together with bits of the form $g_{\sigma_i}(s_{\sigma_i}^{new})$ to make a new G' satisfying $\bigwedge_{i=1}^{k} s_{\sigma_i} \neq a_i$. Since Q_1 does not contain any of the bits we changed, we have $Q_1(G) = Q_1(G')$. Since G' satisfies the σ_i constraints, we have $Q(G') = 0$. Furthermore, if S is any subset of variables of the form $g_{\rho_j}(b_j)$, changing G' on any subset of those variables keeps it in $DRFS_{n,h}$ and does not change any of the values of s_{σ_i}. So for each such S we have $Q(G'^S) = 0$. Applying the reduction lemma, we have $Q_1(G') = 0$ and so $Q_1(G) = 0$ as well. Thus Q_1 satisfies condition 3 for $k - 1$.

By the induction hypothesis, Q_1 has degree at least $k - 1$ and thus Q has degree at least k, completing the proof.

5.4   Wrap-Up

Now we complete the proof of our main result: the degree lower bound of $h + 1$ for any polynomial weakly-representing $RFS_{n,h}$ as stated in Theorem 2.

Proof. Suppose that P is any polynomial that weakly b-represents $RFS_{n,h}$ (where b is either 0 or 1). Since reducing a polynomial on a domain can only decrease its degree, and does not change representation on that domain, we may assume WLOG that P is reduced on $DRFS_{n,h}$. Let X be a $(1 - b)$-input to $RFS_{n,h}$ such that $P(X) \neq 0$. From the definition of $RFS_{n,h}$, the value of $RFS_{n,h}(X)$ changes to b if we change the value of $g_\tau^X(s_\tau^X)$ in X. Hence P must contain the variable $g_\tau(s_\tau^X)$. Let M be a maximal monomial occurring on P containing only variables of the form $g_\tau(a)$ for some $a \in \{0,1\}^n$. If such a monomial M of length $h + 1$ or higher exists, the theorem is proven and we are done. So assume a maximal monomial $M = \prod_{j=1}^k g_\tau(a_j)$ occurs in P and that $1 \leq k \leq h < 2^n$. Factor P as $QM + R$, and consider the total coefficient Q of M in P. We claim that Q satisfies conditions 1, 2, and 3 from the Main Lemma (Lemma 3) for $k = h$.

1. For condition 1, we have that Q is reduced and non-zero on $DRFS_{n,h}$ by the Reduction Lemma.
2. For condition 2, no variable of the form $g_\tau(x)$ occurs in Q, because we factored out a maximal monomial M from P containing variables of this type; and Q was its quotient.
3. For condition 3, we just need to show that $Q(G) = 0$ whenever $G \in DRFS_{n,h}$ and G satisfies $\bigwedge_{i=1}^k (s_\tau^G \neq a_i)$.

   So let G be a valid input that satisfies $\bigwedge_{i=1}^k (s_\tau^G \neq a_i)$, and consider the task of showing $Q(G) = 0$. The value of $RFS_{n,h}(G)$ depends on the bit $g_\tau^G(s_\tau^G)$, but Q does not contain the variable $g_\tau(s_\tau^G)$. So we may as well work with a G' satisfying $RFS_{n,h}(G') = b$, and agreeing with G except possibly at the bit $g_\tau(s_\tau^G)$. This reduces the task to an equivalent one of showing $Q(G') = 0$.

   We will show $Q(G') = 0$ by using the Invariance Lemma (Lemma 2). Applying this lemma requires showing that $P(G') = P(G'^S)$ for every subset S of $\{g_\tau(a_1), \ldots, g_\tau(a_k)\}$. Since G' satisfies $\bigwedge_{i=1}^k (s_\tau^{G'} \neq a_i)$, none of the bits in S affect the recursive Fourier sampling promises of G' or the answer $RFS_{n,h}(G')$. Thus $RFS_{n,h}(G') = RFS_{n,h}(G'^S) = b$ for every subset S of $\{g_\tau(a_1), \ldots, g_\tau(a_k)\}$. Then since P weakly b-represents $RFS_{n,h}$, we have $P(G') = P(G'^S) = b$ for every subset S of $\{g_\tau(a_1), \ldots, g_\tau(a_k)\}$ as well. Now applying the Invariance Lemma, we obtain $Q(G') = 0$. This completes the proof of condition 3 of the Main Lemma for Q.

Applying the Main Lemma, the polynomial Q has degree at least h, and since Q was obtained from P by factoring out a monomial of length at least 1, our weakly b-representing polynomial P has degree at least $h + 1$.

6   Conclusion

Because the polynomial degree of the recursive Fourier sampling problem is low, we cannot directly apply the usual technical tools to resolve its associated longstanding conjectures. As an upside to this though, the degree is so low that we can characterize, through recursive bit manipulations, the problem's "weak" polynomial degree degree right up to the exact integer. Of course, many interesting open questions remain.

References

1. Bernstein, E., Vazirani, U.: Quantum complexity theory. In: STOC 1993: Proceedings of the Twenty-Fifth Annual ACM symposium on Theory of computing, pp. 11–20. ACM Press, New York (1993)
2. Bernstein, E., Vazirani, U.: Quantum complexity theory. SIAM J. Comput. 26, 1411–1473 (1997)
3. Shor, P.W.: Polynomial-time algorithms for prime factorization and discrete logarithms on a quantum computer. SIAM Journal on Computing 26, 1484–1509 (1997)
4. Beals, R., Buhrman, H., Cleve, R., Mosca, M., de Wolf, R.: Quantum lower bounds by polynomials. In: IEEE Symposium on Foundations of Computer Science, pp. 352–361 (1998)
5. Aaronson, S.: Quantum lower bound for the collision problem. In: STOC 2002: Proceedings of the Thiry-Fourth Annual ACM symposium on Theory of Computing, pp. 635–642. ACM, New York (2002)
6. Aaronson, S.: Quantum lower bound for recursive fourier sampling. Quantum Information and Computation 3, 2–72 (2003)
7. Aaronson, S.: Bqp and the polynomial hierarchy. Technical Report ECCC TR09-104 (2009)
8. Furst, M.L., Saxe, J.B., Sipser, M.: Parity, circuits, and the polynomial-time hierarchy. Mathematical Systems Theory 17, 13–27 (1984)
9. Johnson, B.E.: Upper and Lower Bounds for Recursive Fourier Sampling. PhD thesis, University of California at Berkeley, Berkeley, CA, USA, chair - Leo Harrington (2008)

Generalized Self-testing and the Security of the 6-State Protocol

Matthew McKague[1] and Michele Mosca[1,2]

[1] Institute for Quantum Computing and Department of Combinatorics,
University of Waterloo, Waterloo, Ontario N2L 3G1, Canada
[2] Perimeter Institute for Theoretical Physics, 31 Caroline Street North,
Waterloo, ON, N2L 2Y5, Canada

Abstract. Self-tested quantum information processing provides a means for doing useful information processing with untrusted quantum apparatus. Previous work was limited to performing computations and protocols in real Hilbert spaces, which is not a serious obstacle if one is only interested in final measurement statistics being correct (for example, getting the correct factors of a large number after running Shor's factoring algorithm). This limitation was shown by McKague et al. to be fundamental, since there is no way to experimentally distinguish any quantum experiment from a special simulation using states and operators with only real coefficients.

In this paper, we show that one can still do a meaningful self-test of quantum apparatus with complex amplitudes. In particular, we define a family of simulations of quantum experiments, based on complex conjugation, with two interesting properties. First, we are able to define a self-test which may be passed only by states and operators that are equivalent to simulations within the family. This extends work of Mayers and Yao and Magniez et al. in self-testing of quantum apparatus, and includes a complex measurement. Second, any of the simulations in the family may be used to implement a secure 6-state QKD protocol, which was previously not known to be implementable in a self-tested framework.

1 Introduction

In [MY04], [MY98], Mayers and Yao introduced the concept of self-testing quantum apparatus with a test for EPR sources and a select set of measurements. In a parallel development, van Dam et al. [vMMS00] introduced the notion of self-testers for quantum circuits in the case where the dimension of the Hilbert space is known. These results were then combined and improved upon by Magniez et al. in [MMMO06], who give a construction for self-testable circuits without knowledge of the dimension of the Hilbert space.

The Mayers-Yao test, and the test of Magniez et al., only allowed for the testing of states and operators that are equivalent to states and operators in a real Hilbert space. McKague et al. [MMG09] showed that in such settings with untrusted apparatus, one cannot experimentally distinguish a quantum

W. van Dam et al. (Eds.): TQC 2010, LNCS 6519, pp. 113–130, 2011.

system with states and evolution involving complex amplitudes from a special simulation using only real amplitudes. In addition to the implications for self-testing untrusted quantum apparatus, this also resolved an open question posed by Gisin [Gis07] related to the violation of Bell inequalities. It is important to note that the real simulation does not preserve inner product relationships from the system it is simulating. At first glance, this suggests that the well-known 6-state protocol [BBBW84], [Bru98] might not be secure in a setting with untrusted apparatus, since the simulated versions of the six quantum states could be more distinguishable than the proofs of security assume, and an adversary could exploit this additional distinguishability and compromise security.

In fact, it is easy to describe such an insecure implementation of the 6-state protocol with untrusted apparatus, however even an implementation of standard BB84 quantum key establishment with untrusted apparatus is insecure if proper measures are not taken in order to rule out "side-channel" attacks. We show that with comparable precautions as those proposed by Mayers and Yao the 6-state protocol remains secure.

This paper starts by describing a general family of simulations that will reproduce the same statistics as any given "reference" experiment, and are thus experimentally indistinguishable from said experiment. We show how the real Hilbert space simulation given in [MMG09] is equivalent to a special case of this family of simulations. The fact that these simulations work is not very surprising: they are essentially mixtures of the reference experiment, or the complex conjugate of the reference experiment. Thus, we have a more general collection of experiments that are experimentally indistinguishable in a self-testing framework. What is particularly remarkable is that we are able to describe, in section 4.1, a family of self-tests which can only be passed by simulations from the general family we describe (up to equivalence, as defined below). This is summarized in Theorem 2.

The self-tests allow us to put a physical experiment in a general collection of experiments, and we are then able to show that the 6-state protocol is secure for all the experiments within this collection. This shows that it is possible to define a secure 6-state protocol within the self-testing framework.

In section 3, we prepare for the proof of Theorem 2, by discussing the Mayers-Yao self-tested source result given in Theorem 1, a new proof of which is given in appendix C. This new proof is shorter and simpler, and more easily extended to prove our more general result.

Then, in section 4.1, we describe a new self-test for an EPR source and local measurement apparatus that will uniquely characterize the general equivalence class associated with this quantum state and measurement operators.

In section 5, we discuss the cryptographic implications, and why a properly self-tested 6-state protocol is still secure.

Lastly, in section 6, we discuss some open problems and future directions, including the robustness of the generalized self-tests.

2   Simulations

In this section we extend the work of McKague et al. in [MMG09]. There the authors gave a construction that allowed the outcomes of any experiment (the *reference* experiment) to be duplicated (*simulated*) by another experiment (the *simulation*) which is described entirely using real numbers. That is to say, all the states, measurement operators, unitaries, Kraus operators, and Hamiltonians are given as vectors and matrices over the real numbers. Of particular interest here is the fact that the simulation is, in general, not equivalent to the reference experiment according to definition 1 below.

In this section we give a construction for a wider family of simulations. The different simulations in the family are, in general, not equivalent to either the reference experiment nor each other. We will be most interested in states and measurements, but, as with the real simulation in [MMG09], it is also possible to simulate discrete and continuous time evolution.

The simulations rely on the simple observation that transforming an experiment by complex conjugation does not alter the statistics it generates. We could also take a classical mixture of the reference experiment and its complex conjugation, flipping a coin (or controlling on a qubit) beforehand to decide which one to perform. In the remainder of this section we fill in some details about the simulations defined by these mixtures.

2.1   States and Measurements

Consider a reference state[1] $|\psi\rangle$ measured according to a reference POVM $\{P_k\}$. We may duplicate the statistics of this experiment using the complex conjugate state $|\psi^*\rangle$ and POVM $\{P_k^*\}$. In addition, we could do some combination of the two; we may add an additional qubit register which records which of the two experiments to perform: $|0\rangle$ for the reference experiment, and $|1\rangle$ for the complex conjugate. This qubit may be in any state, and not necessarily pure. We then arrive at a new state

$$\rho' = a\,|0\rangle\langle 0|\otimes|\psi\rangle\langle\psi| + (1-a)\,|1\rangle\langle 1|\otimes|\psi^*\rangle\langle\psi^*| + c\,|0\rangle\langle 1|\otimes|\psi\rangle\langle\psi^*| + c^*\,|1\rangle\langle 0|\otimes|\psi^*\rangle\langle\psi| \tag{1}$$

with $a \geq 0$ and $|c| \leq \sqrt{a(1-a)}$. The important feature is that when we project onto $|0\rangle\langle 0|$ or $|1\rangle\langle 1|$ we get either $|\psi\rangle$ or $|\psi^*\rangle$, respectively. For the measurement, we form the POVM

$$\{|0\rangle\langle 0|\otimes P_k + |1\rangle\langle 1|\otimes P_k^*\}. \tag{2}$$

This POVM measurement is equivalent to measuring the added qubit, collapsing the state into either $|\psi\rangle$ or $|\psi^*\rangle$ and then measuring either $\{P_k\}$ or $\{P_k^*\}$ as appropriate; thus the statistics of the experiment are preserved.

Different simulations are arrived at by choosing different values of a and c. If $a = 1$ and $c = 0$ then we obtain the reference experiment. For $a = 0$ and $c = 0$ we obtain the complex conjugate. Another interesting case is when $a = c = \frac{1}{2}$, in which case we obtain (up to a local change of bases) the real simulation of [MMG09] as shown in appendix B.

---

[1] We may consider mixed states as well, but it is not necessary for our discussion.

2.2   Operators

Although it will not be necessary for our discussion, it is possible to simulate a reference experiment which includes evolution, according to a unitary, completely-positive map, or Hamiltonian. The details are discussed in appendix A.

2.3   Non-local Computations

For multi-party experiments, such as the Mayers-Yao test, we would need the simulation to be performed in a local fashion with the measurements operating on local systems only. As defined above this not the case, but it is easy to modify the operators to make it so. We simply add an extra qubit for each party and record in each qubit whether to perform the reference experiment or the complex conjugate. We arrive at states analogous to that in equation 1, but with $|0\rangle$ and $|1\rangle$ replaced with logical states $|\overline{0}\rangle = |00\ldots0\rangle$, $|\overline{1}\rangle = |11\ldots1\rangle$ defined on the extra qubits held by the various parties. Finally, each party conditions their operations on their local copy of the qubit, applying either the reference operation or the complex conjugate.

3   Mayers-Yao Self-test

The goal of the Mayers-Yao test is to compare two experiments. The first experiment is the *reference* experiment, which is the experiment we wish to implement. It is a blueprint, or gold standard, against which we compare the other experiment, the physical experiment, which is the experiment that is actually performed. Within the physical experiment we consider the entire physical apparatus, including the environment, so that we obtain a pure state on a Hilbert space of unknown dimension (however, we will limit ourselves to finite dimensions.) The reference and physical experiments consist of reference and physical states, operations, and measurements. The two experiments are compared through the statistics that they generate.

3.1   Equivalence

The proof considers a particular reference experiment, as described in section 3.2. This experiment is defined on a pair of qubits, so we will limit our discussion to such systems. As well, we consider only pure states - the physical system is unlimited (but finite) in size, so we may include the environment to obtain a pure state. The conclusion of Mayers and Yao is that if the statistics of a physical experiment agree with that of the reference experiment, then the physical experiment is equivalent to the reference experiment, under a particular notion of equivalence.

When defining a notion of equivalence in this setting we must first consider how me might change the reference experiment in a way that preserves the statistics of the outcomes. Any such change is invisible from the perspective of the statistics and hence we cannot rule them out. Here is a list of such changes:

1. Local changes of basis
2. Adding ancillae to physical systems, prepared in any joint state (the measurement does not act on them)
3. Changing the action of the observables outside the support of the state
4. Locally embedding the state and operators in a larger (or smaller) Hilbert space.

In order to accommodate these various changes we define equivalence as follows.

Definition 1. *A reference experiment is described by a n-partite state $|\psi\rangle$ on Hilbert space $\mathcal{X} = \mathcal{X}_1 \otimes \ldots \mathcal{X}_n$ and local measurement observables M_m for various m. Further, consider a physical experiment described by a n-partite state $|\psi'\rangle$ on Hilbert space $\mathcal{Y} = \mathcal{Y}_1 \otimes \cdots \otimes \mathcal{Y}_n$ and local measurement observables M'_m for various m. We say that the physical experiment is* equivalent[2] *to the reference experiment (and the physical state and measurement observables are equivalent to the reference state and measurement observables) if there exists a local isometry*

$$\Phi = \Phi_1 \otimes \ldots \Phi_n, \quad \Phi_j : \mathcal{Y}_j \mapsto \mathcal{Y}_j \otimes \mathcal{X}_j \tag{3}$$

such that

$$\Phi(|\psi'\rangle) = |junk\rangle_\mathcal{Y} \otimes |\psi\rangle_\mathcal{X} \tag{4}$$
$$\Phi(M'_m |\psi'\rangle) = |junk\rangle_\mathcal{Y} \otimes M_m |\psi\rangle_\mathcal{X} . \tag{5}$$

The isometry Φ may be constructed by attaching ancillae in some product state $|00\ldots0\rangle_\mathcal{X}$ and applying local unitaries to the subsystems. Note that if we make any finite number of changes from the list above then we may construct a suitable local isometry and show that the experiment is equivalent to the reference experiment. Also, any experiment that is equivalent to the reference experiment may be constructed by applying changes from the list above: one simply attaches ancillae in the state $|junk\rangle$ and performs a suitable change of basis. The content of the main theorem is that, for a carefully chosen experiment, these are the *only* changes that preserve the statistics.

Theorem 1 (Mayers and Yao). *Suppose a physical experiment reproduces the statistics of the reference experiment described in section 3.2. Then the physical experiment is equivalent to the reference experiment.*

A simplified proof for the Mayers-Yao self-test is give in appendix C.

3.2   Mayers-Yao Self-test Reference Experiment

A general schematic for the Mayers-Yao reference experiment is shown in figure 1. A bipartite state $|\psi\rangle$ is distributed to a pair of measurement devices. The two measurement devices take classical inputs a and b, which each take one of three values. The devices then output classical bits, x and y.

---
[2] Note that this is not an equivalence relation since it is not symmetric.

a

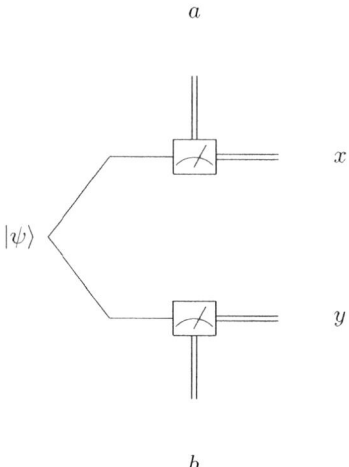

x

$|\psi\rangle$

y

b

Fig. 1. Mayers-Yao self-test circuit

The reference state is an EPR pair $|\phi_+\rangle = \frac{1}{\sqrt{2}}(|00\rangle + |11\rangle)$ and the reference measurement observables are $X, Z, \frac{X+Z}{\sqrt{2}}$ for each side of the EPR pair. For brevity we label $\frac{X+Z}{\sqrt{2}} = D$. For the untrusted physical devices this equality is not given, so there the separate label D is required.

4 Extending the Mayers and Yao Self-test

The original Mayers and Yao EPR test utilized only a small set of measurements. Conspicuously missing is anything with complex coefficients. An important consequence of this is that the circuit test developed by Magniez et al. [MMMO06] is not able to test gates with complex coefficients; only gates with real coefficients can be tested. More specifically, real measurements reveal no information about the imaginary component of a density matrix.

In fact the Mayers-Yao self-test cannot be directly extended to include any measurements with complex coefficients. This is a result of the notion of equivalence used. Suppose that we wish to include the Y measurement in the set of reference measurements. The devices could instead implement $-Y$, the complex conjugate. So long as all complex measurements were complex conjugated it would be impossible to detect this change. Although this does not present an immediate problem - such a transformation is internally consistent and produces the correct outcome statistics - we cannot transform such a circuit back into the reference circuit using unitary transformations.

If this were the whole story we could simply require that the physical circuit be transformable into either the reference circuit or its complex conjugate. However, the real simulation, and now the general family of simulations, are also indistinguishable from the reference circuit and not unitarily transformable into the reference circuit.

We have one encouraging fact: all of the known simulations are equivalent to a simulation from the general family of simulations. We now prove that we can extend the Mayers-Yao test such that these are the only simulations. Hence we may extend our notion of equivalence to include these simulations and obtain a new self-testing theorem.

Theorem 2. *Suppose a physical experiment duplicates the statistics generated by the reference experiment described in section 4.1. Then the physical experiment is equivalent to one of the simulations of the reference experiment described in section 2.*

4.1 Extended Mayers-Yao Self-test Reference Experiment

The extended Mayers-Yao test will consist of three regular Mayers-Yao tests, performed together. Alice and Bob will perform the Mayers-Yao test with measurement settings (labelled with subscript A when used by Alice, and subscript B when used by Bob):

1. X, Z, and D
2. X, Y, and E
3. Y, Z, and F.

In the reference experiment the measurement settings X, Y and Z are realized by the Pauli operators, with $Y_B = -Y$ and otherwise $X_A = X_B = X$, $Y_A = Y$, $Z_A = Z_B = Z$. The other settings are realized by $D_A = \frac{X+Z}{\sqrt{2}}$, $E_A = \frac{X+Y}{\sqrt{2}}$, $F_A = \frac{Y+Z}{\sqrt{2}}$ on Alice's side and $D_B = \frac{X+Z}{\sqrt{2}}$, $E_B = \frac{X-Y}{\sqrt{2}}$, $F_B = \frac{Z-Y}{\sqrt{2}}$ on Bob's side. Bob's Y_B measurements all carry the -1 phase since measuring the state $|\phi_+\rangle$ with the operator $Y \otimes Y$ produces -1 instead of 1 as in the Mayers-Yao reference experiment. The reference state is again $|\phi_+\rangle$.

4.2 Proof of Theorem 2

We start by assuming that the states are all pure as in the Mayers-Yao test. Again we may incorporate the purification of a mixed state into either Alice or Bob's state by adding an ancilla.

First we apply the Mayers-Yao result with the measurements X, Z and D. We find a local isometry Φ as in definition 1. We extend Φ by adding an extra qubit for each side initialized in the state $|0\rangle$. Then Φ takes the X_A, Z_A, X_B and Z_B measurements to $X_{Q_A} \otimes I_{R_A} \otimes I_{S_A}$, $Z_{Q_B} \otimes I_{R_A} \otimes I_{S_A}$, $X_{Q_B} \otimes I_{R_B} \otimes I_{S_B}$ and $Z_{Q_B} \otimes I_{R_B} \otimes I_{S_B}$ where R_A and R_B are the added qubit registers and S_A and S_B are the junk registers. Meanwhile the state has the form $|\phi_+\rangle_{Q_A Q_B} \otimes |00\rangle_{R_A R_B} \otimes |junk\rangle_{S_A S_B}$.

We now consider the remaining measurements. The reference experiments for these measurements can be transformed using local unitaries into the usual Mayers-Yao reference experiments. Thus we may apply the result. However, we stop short of using the full result. Within the proof of Theorem 1 we achieve the following result.

Lemma 1. *Suppose a physical experiment reproduces the statistics of the Mayers-Yao reference experiment described in section 3.2. Then the physical measurements X_A and Z_A anti-commute on the support of the physical state, as do X_B and Z_B.*

This is shown in section C.2. When we apply this result to the remaining measurements in the extended test, we find that X_A and Y_A anti-commute on the support of the state, as do X_B and Y_B, Z_A and Y_A and Z_B and Y_B. For the remaining discussion we will limit ourselves to the support of the state.

Consider the A side measurements first. We may express Y_A as

$$Y_A = \sum_{P,E} y_{P,E} P_{Q_A} \otimes I_{R_A} \otimes E_{S_A}$$

where the P ranges over the Pauli operators and the E ranges over a basis for the Hermitian operators on S_A.

Since Y_A anti-commutes with $X_{Q_A} \otimes I_{R_A S_A}$ the coefficients of all the terms with $P = X$ must be 0. Indeed, since $-Y_A = (X_{Q_A} \otimes I_{R_A S_A}) Y_A (X_{Q_A} \otimes I_{R_A S_A})$ we have

$$-\sum_{P,E} y_{P,E} P_{Q_A} \otimes I_{R_A} \otimes E_{S_A} = \sum_{P \in \{I,X\},E} y_{P,E} P_{Q_A} \otimes I_{R_A} \otimes E_{S_A} - \sum_{P \in \{Y,Z\},E} y_{P,E} P_{Q_A} \otimes I_{R_A} \otimes E_{S_A}$$

where on the right hand side we have separated out the terms that commute with $X_{Q_A} \otimes I_{R_A S_A}$ and those that anti-commute. We see that we must have $y_{X,E} = -y_{X,E} = 0$ and $y_{I,E} = -y_{I,E} = 0$ for all E.

Applying similar reasoning and the test with Y and Z we find that $y_{Z,E} = 0$ for all E. Thus $Y_A = Y_{Q_A} \otimes I_{R_A} \otimes M_{S_A}$ for some Hermitian and unitary M_{S_A}. Next we compose Φ with a "phase kickback" circuit consisting of a Hadamard gate on the R_A register, followed by a controlled M_{S_A}, controlled on the R_A register, and a final Hadamard gate on the R_A register. This results in a new isometry (we will still call it Φ) such that

$$\Phi(Y_{Q_A} \otimes M_{S_A} |\psi\rangle) = Y_{Q_A} \otimes Z_{R_A} |\phi_+\rangle_Q |junk\rangle_{RS} . \tag{6}$$

This is essentially the well known translation of a two outcome measurement into a qubit measurement. Also, since the addition of the phase kickback did not operate on the junk register the X and Z measurements are not affected.

The above process can be repeated for Bob's side, with analogous conclusions. In order to be consistent with the reference experiment, we may construct our isomorphism so that

$$\Phi(Y_{Q_B} \otimes M_{S_B} |\psi\rangle) = Y_{Q_B} \otimes Z_{R_B} |\phi_+\rangle_Q |junk\rangle_{RS} . \tag{7}$$

We have thus shown that the measurements are as in the general simulation.

We now turn our attention to the state. From the Mayers-Yao test on X and Z we know that the state on $Q_A \otimes Q_B$ (after applying Φ) is $|\phi_+\rangle$. We next consider the state on the remaining registers, $|junk\rangle_{RS}$. We may express this in the singular value (Schmidt) decomposition, split between R_{AB} and S_{AB}:

$$|\theta\rangle = \sum_j \lambda_j |j\rangle_{R_{AB}} |j\rangle_{S_{AB}} \tag{8}$$

with $\lambda_j > 0$. Since the Y measurement setting gives correlated results (recall we introduced a -1 factor on the B side measurement observable) and the form of Y_A and Y_B, the states $|j\rangle_{R_{AB}}$ must all be +1 eigenvectors of $Z_{R_A} \otimes Z_{R_B}$. If this were not the case then a -1 phase would be introduced and the measurement results would be incorrect at least some of the time. Thus the only possible states for $|j\rangle_{R_{AB}}$ are superpositions of $|00\rangle$ and $|11\rangle$. We do some relabelling and arrive at

$$|\psi\rangle = |\phi_+\rangle_{Q_{AB}} \otimes \left(\alpha |00\rangle_{R_{AB}} |\theta_{00}\rangle_{S_{AB}} + \beta |11\rangle_{R_{AB}} |\theta_{11}\rangle_{S_{AB}}\right) \qquad (9)$$

with $|\theta_{00}\rangle$ and $|\theta_{11}\rangle$ not necessarily orthogonal. Note that tracing out the S_{AB} ancillae results in a state exactly as described by the multi-party simulation in section 2. Thus we have demonstrated that the physical experiment is equivalent to one of the general simulations of the reference experiment, and completed the proof of Theorem 2.

5   Cryptographic Setting

Suppose that two or more parties are engaged in a cryptographic protocol using self-tested apparatus. The extended Mayers-Yao test above allows them to determine that the devices are implementing a simulation from the family of simulations described in section 2. Suppose further that the adversary, Eve, knows how the devices are implemented and controls the preparation of the state. The honest parties only perform operations as specified for the simulation. Eve, on the other hand, is free to interact with the extra qubits in the simulation in any way she likes. Does this give any advantage to Eve?

Eve can potentially perform many operations, including entangling a qubit of her own with the extra simulation qubits allowing her to perform simulation operations. She may also interact in complex ways with the extra simulation qubits along with the original register. Despite this, we are able to prove that Eve can gain no advantage for some protocols.

We explore a restricted class of protocols that are especially easy to analyse. These are protocols where the only operation that an honest party will do is a Pauli measurement. This class includes the six-state quantum key distribution protocol (implemented in as an entanglement based protocol) [BBBW84], [Bru98]. We will demonstrate that these protocols do not leak any more information when implemented using one of the simulations.

The proof is a series of security reductions to protocols in which each reduction only increases Eve's power. We will show that the final protocol in the reduction is just as secure as the reference protocol (without the simulation applied), hence the simulation protocol is also just as secure as the reference protocol.

For the first reduction we suppose that the participants in the protocol measure their simulation qubit in the Z eigenbasis after the protocol is completed, and transmit the result to Eve. This does not interfere with the intended protocol and only increases Eve's information. Since the Z measurement commutes with all simulation operations, the participants could just as well have performed the measurement before the protocol began. If Eve is the one who prepares the initial state for the simulation (in other cases Eve has strictly less power) then Eve could also perform this measurement herself. This measurement would collapse the state to an eigenvector of the Z measurements and Eve's strategy would be a mixture of different strategies with the states each an eigenvector of the Z measurements.

Let us examine the result of Eve choosing one of these eigenvector states. Each of the parties will receive their extra qubit prepared in a Z eigenvector. The effect of this on their operations is either to perform the protocol's original operation (in the case of a $|0\rangle$) or the complex conjugate (in the case of a $|1\rangle$.) For Pauli measurements, only the Y measurement is affected: the output bit is flipped in the case of the complex conjugate.

If every party receives the same eigenvector in their extra qubit, then the protocol reduces to either the original or the complex conjugate. In either case the security is identical to the original protocol. If the extra qubits are not in the same eigenvector then some Y measurements outcomes will be flipped and some will not. This does not affect Eve's information since she controls which outcomes are flipped and can undo the flips in her reckoning of the final classical information. Note that the bit flips may introduce errors into the protocol. If the protocol does not explicitly check for such errors (as does the 6-state protocol) information will still not be leaked to Eve, however a test for these errors may be required to make sure the protocol functions correctly. The final protocol, and hence the simulation, is thus as secure as the original protocol.

6   Conclusions and Future Work

6.1   Conclusions

Theorem 2, along with the security result of section 5, allows us to analyze the case of the 6-state QKD protocol in the self-tested framework. In particular we may define a self-testing version of the 6-state protocol in which the extended Mayers-Yao test is incorporated along with the usual 6-state protocol. Given a robust version of the test (see section 6.2) we may first estimate the state and measurement observables, then apply a security proof for the 6-state protocol in order to derive a secure key rate.

Although a self-tested 6-state protocol is currently not practical, nor likely to become so, the result is interesting from a theoretical perspective within the self-tested framework. Previous results were limited to real Hilbert spaces, one could apply the real simulation explicitly within the reference experiment and then proceed with the self-test. This works fine for circuits, where only the correct outcome is important, however the 6-state protocol introduces other concerns, namely the possibility of information leaking to an adversary. The current work thus illustrates how a self-test for complex operations provides additional benefit over the previous self-tests.

6.2   Future Work

Note that we have not described a physically realizable test in section 4. The proof requires that the expected value of the observables match the reference exactly. This cannot be established physically without some kind of repeatability assumptions and an infinite number of trials. The original test by Mayers and Yao was shown to be robust in [MMMO06], establishing a polynomial relationship between the precision of the statistics and the closeness to an EPR state. We are currently studying the robustness of these new tests. This is an important line of future research. A related task is to extend the results to continuous variable systems.

Another interesting line of research is to follow the same path as Magniez et al. to obtain a self-testing circuit for arbitrary circuits, now allowing complex gates. The framework and proofs from [MMMO06] offer a roadmap for such research, but there are some technical problems that arise along the way so a straightforward adaptation is not possible. These are due to the larger Hilbert space created when adding the extra qubits to allow the simulations.

Acknowledgements. This work is supported by Canada's NSERC, Quantum-Works, Ontario Centres of Excellence, MITACS, CIFAR, CRC, ORF, the Government of Canada, and Ontario-MRI.

References

[BBBW84]   Bennett, C.H., Brassard, G., Breidbart, S., Wiesner, S.: Eavesdrop-detecting quantum communications channel. IBM Technical Disclosure Bulletin 26(8), 4363–4366 (1984)

[Bru98]   Bruß, D.: Optimal eavesdropping in quantum cryptography with six states. Phys. Rev. Lett. 81(14), 3018–3021 (1998), doi:10.1103/PhysRevLett.81.3018

[Gis07]   Gisin, N.: Bell inequalities: many questions, a few answers (2007)

[MMG09]   McKague, M., Mosca, M., Gisin, N.: Simulating quantum systems using real Hilbert spaces. Physical Review Letters 102(2), 20505 (2009), http://link.aps.org/abstract/PRL/v102/e020505, doi:10.1103/PhysRevLett.102.020505

[MMMO06]   Magniez, F., Mayers, D., Mosca, M., Ollivier, H.: Self-testing of quantum circuits. In: Bugliesi, M., Preneel, B., Sassone, V., Wegener, I. (eds.) ICALP 2006. LNCS, vol. 4052, pp. 72–83. Springer, Heidelberg (2006)

[MY98]   Mayers, D., Yao, A.: Quantum cryptography with imperfect apparatus. In: FOCS, pp. 503–509 (September 1998), http://arxiv.org/abs/quant-ph/9809039

[MY04]   Mayers, D., Yao, A.: Self testing quantum apparatus. QIC 4(4), 273–286 (2004), http://arxiv.org/abs/quant-ph/0307205

[vMMS00]   van Dam, W., Magniez, F., Mosca, M., Santha, M.: Self-testing of universal and fault-tolerant sets of quantum gates. In: STOC 2000: Proceedings of the Thirty-Second Annual ACM Symposium on Theory of Computing, pp. 688–696. ACM, New York (2000), doi:10.1145/335305.335402

A   Evolution in Simulations

We can extend the measurement operator defined in 2 to arbitrary operators. We define

$$C(M) = |0\rangle\langle 0| \otimes M + |1\rangle\langle 1| \otimes M^*. \qquad (10)$$

Note that $C(M)$ can be expressed differently as

$$C(M) = I \otimes Re(M) + iZ \otimes Im(M) \qquad (11)$$

where $Re(M)$ and $Im(M)$ are the real and imaginary parts of M (both real matrices). In the case of a multi-party simulation, the Z operates on a particular party's added qubit.

We summarize some of the properties of $C(M)$ here

Lemma 2. *Let M and N be matrices. Then we have the following:*

1. $C(MN) = C(M)C(N)$.
2. $C(M + N) = C(M) + C(N)$.
3. *Let a be a real number, then $C(aM) = aC(M)$.*
4. *If $|\psi\rangle$ is an eigenvector of M with eigenvalue λ, then $|0\rangle|\psi\rangle$ and $|1\rangle|\psi\rangle$ are eigenvectors of $C(M)$ with eigenvalues λ and λ^*, respectively.*
5. $C(M)$ *is Hermitian if and only if M is.*
6. $C(M)$ *is unitary if and only if M is.*
7. $C(M)$ *is positive semi-definite if and only if M is.*
8. *When M is Hermitian, $Tr(C(M)) = 2\,Tr(M)$.*

These properties can be derived easily.

Discrete time evolution. The properties of $C(\cdot)$ allow us to easily determine how the simulation states in the continuum evolve. Let U and $|\psi\rangle$ be a reference unitary operation and state and let ρ' be as in equation 1. By the form of $C(U)$ we have

$$C(U)\rho'C(U)^\dagger = a|0\rangle\langle 0| \otimes U|\psi\rangle\langle\psi|U^\dagger + (1-a)|1\rangle\langle 1| \otimes U^*|\psi^*\rangle\langle\psi^*|U^T +$$

$$c|0\rangle\langle 1| \otimes U|\psi\rangle\langle\psi^*|U^T + c^*|1\rangle\langle 0| \otimes U^*|\psi^*\rangle\langle\psi|U^\dagger.$$

But this is the simulation state for $U|\psi\rangle$, and hence $C(U)$ evolves the simulation state ρ' to produce a new simulation state corresponding to $U|\psi\rangle$. Compositions of unitaries will also evolve the state correctly so that the measurement statistics at the end of a circuit will be identical to that of the reference circuit.

General quantum operations may be mapped similarly. It is easy to verify that in Kraus representation a completely positive map is mapped to a completely positive map if we apply $C(\cdot)$ to each of the Kraus operators. The trace preserving property is also preserved. We apply the same reasoning as for U above to with each Kraus operator. The linearity of C then allows us to conclude that the simulation map will behave correctly. That is to say, it will map ρ' to a new simulation state corresponding to $|\psi\rangle$ evolved under the reference map.

Continuous time evolution. We begin with a Hamiltonian H. One can simulate the Schrödinger evolution of H on $|\psi\rangle$ by evolving H^* on $|\psi^*\rangle$ backwards in time, or equivalently, evolving the system according to $-H^*$, and measuring with conjugated observables.

Thus, the simulation of the evolution of H can be achieved using the Hamiltonian

$$H' = |0\rangle\langle0| \otimes H - |1\rangle\langle1| \otimes H^*. \tag{12}$$

The evolution of the state according to the Schrödinger equation

$$U(t) = e^{-iH't} \tag{13}$$

gives

$$e^{-iH't} = |0\rangle\langle0|\otimes e^{-iHt}+|1\rangle\langle1|\otimes e^{-i(-H^*)t} = |0\rangle\langle0|\otimes e^{-iHt}+|1\rangle\langle1|\otimes\left(e^{-iHt}\right)^* = C(e^{-iHt}) \tag{14}$$

(using the fact that $\exp(A + B) = \exp(A) + \exp(B)$ when $AB = 0 = BA$, and that $\exp(P \otimes A) = P \otimes \exp(A)$ when $P^2 = P$). Thus

$$e^{-iH't} = C(e^{-iHt}) \tag{15}$$

and the simulation evolution tracks that of the reference system.

Another way to arrive at the same H' is the approach used in the real simulation [MMG09]. There, rather than considering the Hamiltonian alone, the whole matrix in the exponent, $-iHt$, was considered. Applying $C(\cdot)$ to this matrix we obtain

$$|0\rangle\langle0| \otimes (-iHt) + |1\rangle\langle1| \otimes (-iHt)^* = i\left(|0\rangle\langle0| \otimes H - |1\rangle\langle1| \otimes H^*\right)t \tag{16}$$

Here the fact that $a^*b^* = (ab)^*$ means $(-iH)^* = iH^*$ and the -1 factor is explained.

B   Real Simulation in the Family

The real simulation presented in [MMG09] can be expressed as a simulation in the family defined above through a change of basis. Starting with the state defined as in 1 with $a = c = \frac{1}{2}$ the simulation state is pure and equal to

$$|\psi'\rangle = \frac{1}{\sqrt{2}}|0\rangle|\psi\rangle + \frac{1}{\sqrt{2}}|1\rangle|\psi^*\rangle.$$

We next apply a Hadamard gate followed by the relative phase rotation

$$\begin{pmatrix} 1 & 0 \\ 0 & -i \end{pmatrix}$$

to the extra qubit. This is the same as applying the unitary

$$U = \begin{pmatrix} 1 & 1 \\ -i & i \end{pmatrix}. \tag{17}$$

The resulting state is

$$\frac{1}{2} |0\rangle (|\psi\rangle + |\psi^*\rangle) - \frac{i}{2} |1\rangle (|\psi\rangle - |\psi^*\rangle)$$

which can be rewritten as

$$|0\rangle \, Re(|\psi\rangle) + |1\rangle \, Im(|\psi\rangle)$$

which is the real simulation described in [MMG09][3].

Operators are transformed quite easily. For operator M we conjugate $C(M)$ by $U \otimes I$. From 11 we see that the resulting operator is

$$(U \otimes I)C(M)(U^\dagger \otimes I) = I \otimes Re(M) + XZ \otimes Im(M) \tag{18}$$

which is exactly the operator used in the real simulation for M.

The states used in the multi-party simulation in [MMG09] are stabilized by $Y_s \otimes Y_t$ for distinct s, t. Also note that the states used in the simulations defined here are stabilized by $Z_s \otimes Z_t$ for distinct s, t. The qubit-wise transformation applied transformations Z into Y, so the multi-party states are also transformed correctly.

C Simplified Proof for Mayers-Yao Self-test

C.1 Proof Overview

The main advantages of the following new proof for the Mayers-Yao self-test is that it is shorter, clearer, and more naturally extends to the more general test given in this paper.

The proof has two distinct parts. The first part establishes some equations on the state and observables based on the observed statistics. These are straightforward and are a direct result of the statistics observed. Next we use these equations to show that the X and Z observables on each side anti-commute on the support of the state. The second part uses the anti-commuting observables to construct local isometries that take the state and observables to the reference state and observables.

One important consideration is that of the support of the state. Since we do not make any claims about the state and observables outside the support of the state we disregard the rest of the Hilbert space. In this way we will not make any more reference to the support of the state.

---

[3] This part of the real simulation was previously well known.

C.2   Observed Statistics Imply Anti-commuting Observables

Statistics. In the reference test the marginals for each observable are all 0. That is,

$$\langle \phi_+ | M \otimes I | \phi_+ \rangle = 0$$

for $M \in \{X, Z, D\}$. (Swapping the systems in this and the following equations gives the same result since $|\phi_+\rangle$ is symmetric.) Measuring the same observable on both sides always give identical outcomes. Thus

$$\langle \phi_+ | M \otimes M | \phi_+ \rangle = 1.$$

Additionally, X and Z measurements are uncorrelated.

$$\langle \phi_+ | X \otimes Z | \phi_+ \rangle = 0.$$

The interesting part comes when we measure X or Z on one side and D on the other.

$$\langle \phi_+ | X \otimes D | \phi_+ \rangle = \langle \phi_+ | Z \otimes D | \phi_+ \rangle = \frac{1}{\sqrt{2}}.$$

State equalities. Using the equations from section 3.2 on the measurement outcomes combined with the fact that $|\psi\rangle$ is normalized gives us the following equations

$$|\psi\rangle = X_A \otimes X_B |\psi\rangle \tag{19}$$
$$= Z_A \otimes Z_B |\psi\rangle \tag{20}$$
$$= D_A \otimes D_B |\psi\rangle \tag{21}$$
$$X_A \otimes I |\psi\rangle = I \otimes X_B |\psi\rangle \tag{22}$$
$$Z_A \otimes I |\psi\rangle = I \otimes Z_B |\psi\rangle \tag{23}$$
$$D_A \otimes I |\psi\rangle = I \otimes D_B |\psi\rangle \tag{24}$$
$$X_A Z_A \otimes I |\psi\rangle = I \otimes Z_B X_B |\psi\rangle \tag{25}$$
$$Z_A X_A \otimes I |\psi\rangle = I \otimes X_B Z_B |\psi\rangle \tag{26}$$
$$X_A Z_A \otimes I |\psi\rangle = X_A \otimes Z_B |\psi\rangle \tag{27}$$
$$Z_A X_A \otimes I |\psi\rangle = Z_A \otimes X_B |\psi\rangle \tag{28}$$

We can also establish some orthogonality relationships between various vectors. In particular the vectors $|\psi\rangle$, $X_A \otimes I |\psi\rangle$, $Z_A \otimes I |\psi\rangle$, $X_A Z_A \otimes I |\psi\rangle$ are pairwise orthogonal.

Our goal for the remainder of the proof is to show that any state for which these equations hold must be equivalent to $|\phi_+\rangle$.

Anti-commuting observables. We now move to more salient matters. First, we note that $D_A \otimes I |\psi\rangle$ must be in the space spanned by $X_A \otimes I |\psi\rangle$ and

$Z_A \otimes I \,|\psi\rangle$ because it has overlap $\frac{1}{\sqrt{2}}$ with each of these orthogonal vectors, and it has norm 1. Thus

$$D_A \otimes I \,|\psi\rangle = \frac{X_A + Z_A}{\sqrt{2}} \otimes I \,|\psi\rangle$$

and analogously for $I \otimes D_B \,|\psi\rangle$. This allows us to make the following deductions

$$|\psi\rangle = D_A \otimes D_B \,|\psi\rangle$$
$$= \frac{1}{2}(X_A + Z_A) \otimes (X_B + Z_B) \,|\psi\rangle$$
$$= |\psi\rangle + (X_A \otimes Z_B + Z_A \otimes X_B) \,|\psi\rangle$$

Applying equations 22 and 23 we obtain

$$(X_A Z_A + Z_A X_A) \otimes I \,|\psi\rangle = 0. \tag{29}$$

By Lemma 3, below, it follows that X_A and Z_A anti-commute on the support of $|\psi\rangle$ on A. Similarly, the observables X_B and Z_B anti-commute on support of $|\psi\rangle$ on B.

Lemma 3. *Let X_A and Z_A be operators and $|\psi\rangle_{AB}$ a bipartite state such that*

$$X_A Z_A \otimes I_B \,|\psi\rangle_{AB} = -Z_A X_A \otimes I_B \,|\psi\rangle_{AB} \,. \tag{30}$$

then $X_A Z_A \,|\phi\rangle = -Z_A X_A \,|\phi\rangle$ for any $|\phi\rangle$ in the support of $|\psi\rangle_{AB}$ on A.

Proof. Let

$$|\psi\rangle = \sum_j \lambda_j \,|j\rangle_A \,|j\rangle_B \,. \tag{31}$$

be the singular value decomposition of $|\psi\rangle$. We then have

$$X_A Z_A \otimes I_B \sum_j \lambda_j \,|j\rangle_A \,|j\rangle_B = -Z_A X_A \otimes I_B \sum_j \lambda_j \,|j\rangle_A \,|j\rangle_B \,. \tag{32}$$

We now take the inner product with $|k\rangle_A \,|k'\rangle_B$ for some k, k' to obtain

$$\lambda_j \,\langle k|_A X_A Z_A \,|j\rangle_A = -\lambda_j \,\langle k|_A X_A Z_A \,|j\rangle_A \tag{33}$$

When we restrict to the subspace to the subspace spanned by the $|k\rangle_A$ for which $\lambda_k \neq 0$ (i.e. on the support of $|\psi\rangle$ on A) we find that $X_A Z_A = -Z_A X_A$.

C.3   Local Unitary Transformations

Now we can easily build the local unitaries required to extract the EPR pair. We use the circuit shown in figure 2. There the outer $|0\rangle$ states are added while the two inner wires carry the two halves of the bipartite state $|\psi\rangle$. This circuit essentially builds a SWAP gate out of two CNOT gates (the usual third gate is not necessary since we initialize with $|0\rangle$.) The SWAP gate extracts the entanglement out of $|\psi\rangle$ and swaps in a product state.

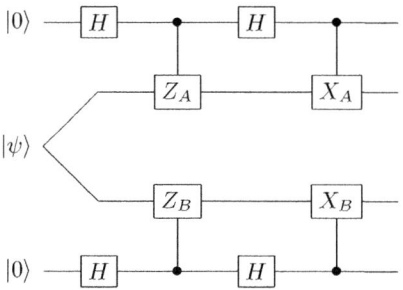

Fig. 2. Circuit for Φ showing equivalence of physical circuit to reference circuit in Mayers-Yao test

The circuit gives two isometries, one for each wire in EPR test circuit, which we denote Φ_A and Φ_B.

State. After applying this circuit the resulting state is

$$\Phi_A \otimes \Phi_B(|\psi\rangle) = \frac{1}{4}(I + Z_A) \otimes (I + Z_B)|\psi\rangle|00\rangle$$
$$+ \frac{1}{4}(I + Z_A) \otimes X_B(I - Z_B)|\psi\rangle|01\rangle$$
$$+ \frac{1}{4}X_A(I - Z_A) \otimes (I + Z_B)|\psi\rangle|10\rangle$$
$$+ \frac{1}{4}X_A(I - Z_A) \otimes X_B(I - Z_B)|\psi\rangle|11\rangle$$

Applying some equations and the anti-commuting result from the previous section we find that this is equal to

$$\Phi_A \otimes \Phi_B(|\psi\rangle) = \frac{1}{4}(I + Z_A) \otimes (I + Z_B)|\psi\rangle(|00\rangle + |11\rangle) +$$
$$(I + Z_A)(I - Z_A) \otimes X_B|\psi\rangle|01\rangle + X_A \otimes (I + Z_B)(I - Z_B)|\psi\rangle|10\rangle$$
$$= \frac{1}{\sqrt{2}}(I \otimes I + I \otimes Z_B)|\psi\rangle|\phi_+\rangle$$

This may look curious since $I + Z_A$ and $I + Z_B$ are not unitary. In fact it is easy to show that the final state still has the correct norm. To give some intuition, note that in the reference case we want to extract $|\phi_+\rangle$ and swap in $|00\rangle = \frac{1}{2\sqrt{2}}(I + Z) \otimes (I + Z)|\phi_+\rangle$.

Measurement operators. We now turn to equivalence of the measurement operators. We start with X_A (the result for X_B follows analogously). Applying X_A to $|\psi\rangle$ before applying the circuit is the same as applying it at the end, with a -1 phase introduced by anti-commuting past the controlled Z_A operation (recall from section C.2 that X_A and Z_A anti-commute on the relevant subspace). The resulting state is

$$\Phi_A \otimes \Phi_B(X_A \otimes I_B \, |\psi\rangle) = \frac{1}{4} X_A(I - Z_A) \otimes (I + Z_B) \, |\psi\rangle \, |00\rangle$$
$$+ \frac{1}{4} X_A(I - Z_A) \otimes X_B(I - Z_B) \, |\psi\rangle \, |01\rangle$$
$$+ \frac{1}{4}(I + Z_A) \otimes (I + Z_B) \, |\psi\rangle \, |10\rangle$$
$$+ \frac{1}{4}(I + Z_A) \otimes X_B(I - Z_B) \, |\psi\rangle \, |11\rangle$$

Following the same logic as used in the state equivalence, we find that the final state is

$$\Phi_A \otimes \Phi_B(X_A \otimes I \, |\psi\rangle) = \frac{1}{\sqrt{2}}(I \otimes I + I \otimes Z_B) \, |\psi\rangle \, (X \otimes I) \, |\phi_+\rangle$$

For the Z_A operation, we see that the effect is a -1 phase kicked back through the final controlled X_A operation. This phase appears on the terms with $|1\rangle$ in the qubit, exactly as if a Z operation had been applied to the qubit. The equivalence for the D operators results from the fact that $D = \frac{X+Z}{\sqrt{2}}$ on the relevant subspace, and linearity.

This concludes the proof of Theorem 1 .

A Conceptually Simple Proof of the Quantum Reverse Shannon Theorem

Mario Berta[1,2], Matthias Christandl[1,2], and Renato Renner[1]

[1] Institute for Theoretical Physics, ETH Zurich, 8093 Zurich, Switzerland
{berta,christandl,renner}@phys.ethz.ch
[2] Faculty of Physics, Ludwig-Maximilians-Universität München, 80333 Munich, Germany

Abstract. The Quantum Reverse Shannon Theorem states that any quantum channel can be simulated by an unlimited amount of shared entanglement and an amount of classical communication equal to the channel's entanglement assisted classical capacity. In this extended abstract, we summarize a new and conceptually simple proof of this theorem [journal reference: arXiv.org:quant-ph/0912.3805], which has previously been proved in [Bennett et al., arXiv.org:quant-ph/0912.5537]. Our proof is based on optimal one-shot Quantum State Merging and the Post-Selection Technique for quantum channels.

1 Introduction

The birth of classical information theory can be dated to 1948, when Shannon derived his famous 'Noisy Channel Coding Theorem' [23]. It shows that the capacity C of a classical channel $\mathcal{J}$ is given by the maximum, over the input distributions X, of the input/output mutual information

$$C = \max_X (H(X) + H(\mathcal{J}(X)) - H(X, \mathcal{J}(X))) .$$

Shannon also showed that the capacity does not increase if one allows to use shared randomness between the sender and the receiver. In 2001 Bennett et al. [3] proved the so called 'Classical Reverse Shannon Theorem' which states that, given free shared randomness between the sender and the receiver, every channel can be simulated using an amount of classical communication equal to the capacity of the channel. This is particularly interesting because it implies that in the presence of free shared randomness, the capacity of a channel $\mathcal{J}$ to simulate another channel $\mathcal{I}$ is given by the ratio of their plain capacities $C_R(\mathcal{J}, \mathcal{I}) = \frac{C(\mathcal{J})}{C(\mathcal{I})}$ and hence only a single parameter remains to characterize classical channels.

In contrast to the classical case, a quantum channel has various distinct capacities [3, 9, 12, 17, 22, 24]. In [3] Bennett et al. argue that the entanglement assisted classical capacity C_E of a quantum channel $\mathcal{E}$ is the natural quantum generalization of the classical capacity of a classical channel. They show that

W. van Dam et al. (Eds.): TQC 2010, LNCS 6519, pp. 131–140, 2011.
© Springer-Verlag Berlin Heidelberg 2011

the entanglement assisted classical capacity is given by the quantum mutual information

$$C_E = \max_\rho (H(\rho) + H(\mathcal{E}(\rho)) - H((\mathcal{E} \otimes \mathrm{id})\Phi_\rho)) \ ,$$

where the maximum goes over all input distributions ρ and Φ_ρ is a purification of ρ. Motivated by this, they conjectured the 'Quantum Reverse Shannon Theorem (QRST)' in [3]. Subsequently Bennett, Devetak, Harrow, Shor and Winter proved the theorem in [2]. The theorem states that any quantum channel can be simulated by an unlimited amount of shared entanglement and an amount of classical communication equal to the channel's entanglement assisted classical capacity. So if entanglement is for free, we can conclude in complete analogy with the classical case, that the capacity of a quantum channel $\mathcal{E}$ to simulate another quantum channel $\mathcal{F}$ is given by $C_E(\mathcal{E}, \mathcal{F}) = \frac{C_E(\mathcal{E})}{C_E(\mathcal{F})}$ and hence only a single parameter remains to characterize quantum channels.

Free entanglement in quantum information theory is usually given in the form of maximally entangled states. But for the Quantum Reverse Shannon Theorem it surprisingly turned out that maximally entangled states are not the appropriate resource for general input sources. More precisely, if one has only maximally entangled states as an entanglement resource, even if these are arbitrarily many, the Quantum Reverse Shannon Theorem cannot be proven [2]. This is because of an issue known as entanglement spread, which basically comes from the fact that entanglement cannot be conditionally discarded without either using communication or causing decoherence [11]. If we change the entanglement resource from maximally entangled states to embezzling states [29] however, the problem of entanglement spread can be overcome.

Definition 1.1 (Hayden, van Dam [29]). *A bipartite state of the form*

$$|\mu(k)\rangle_{AB} = \frac{1}{\sqrt{G(k)}} \sum_{j=1}^{k} \frac{1}{\sqrt{j}} |jj\rangle_{AB} \ , \tag{1}$$

where $G(k) = \sum_{j=1}^{k} \frac{1}{j}$, *is called* embezzling state *of index* k *(which is the Schmidt-rank of* $|\mu(k)\rangle$*).*

These states have the following special feature.

Proposition 1.2 (Hayden, van Dam [29]). *Let* $\epsilon > 0$ *and let* $|\varphi\rangle_{AB}$ *be a bipartite pure (and normalized) state of Schmidt-rank* m*. Then the transformation*

$$|\mu(k)\rangle_{AB} \mapsto |\mu(k)\rangle_{AB} \otimes |\varphi\rangle_{AB} \tag{2}$$

can be accomplished with fidelity[1] better than $(1 - \epsilon)$ *for* $k > m^{1/\epsilon}$ *without any communication.*

---

[1] The fidelity between two normalized states ρ and σ is defined as $F(\rho, \sigma) = \|\sqrt{\rho}\sqrt{\sigma}\|_1$, where $\|\Gamma\|_1 = \mathrm{tr}\sqrt{\Gamma \Gamma^\dagger}$ is the trace norm.

In this extended abstract we summarize a conceptually simple proof of the Quantum Reverse Shannon Theorem. It is based on a new one-shot version for 'Quantum State Merging' and the 'Post-Selection Technique' for quantum channels. As in [2] we make use of embezzling states.

Quantum State Merging is a well known quantum information processing primitive [1, 13, 14, 18] and corresponds to the quantum generalization of classical Slepian and Wolf coding [25]. There are basically two (equivalent) versions of State Merging, one of them also called the 'Mother Protocol' [1]. Let ρ_{AB} be a bipartite quantum state, where A is with a party *Alice* and B is with another party *Bob*. State Merging answers the question of how much of a given resource (classical/quantum communication, entanglement) is needed in order to optimally transfer the A-part of ρ_{AB} to Bob (relative to a purifying system R).[2] The dual of this, called *Quantum State Splitting*, addresses the problem of how much of a given resource (classical/quantum communication, entanglement) is needed in order to transfer the B' part of a state $\rho_{B'B}$ held by Bob, (back) to Alice (relative to a purifying system R).[3]

The *Post-Selection Technique* was introduced in [7] and is a tool to show that two completely positive and trace preserving (CPTP) maps, that act symmetrically on an n-partite system, are almost equal, in the sense that they are close in the metric induced by the diamond norm [15]. The diamond norm involves a maximization over all possible input states. The Post-Selection Technique allows to drop this maximization. In fact, it suffices to consider a single de Finetti type input state (i.e. a state which consist of n identical and independent copies of an (unknown) state on a single subsystem).

Our proof of the Quantum Reverse Shannon Theorem is based on the following idea. Let $\mathcal{E}_{A\rightarrow B}$ be a quantum channel that takes inputs ρ_A on Alice's side and outputs $\mathcal{E}_{A\rightarrow B}(\rho_A)$ on Bob's side. To find a way to simulate this quantum channel, it is useful to think of $\mathcal{E}_{A\rightarrow B}$ as

$$\mathcal{E}_{A\rightarrow B}(\rho_A) = \text{tr}_{A'}(U_{A\rightarrow BA'}\rho_A U^\dagger_{A\rightarrow BA'})\,,$$

where A' is an additional register and $U_{A\rightarrow BA'}$ is some isometry from A to BA'. This is the Stinespring dilation [26]. The idea is to first simulate the quantum channel locally at Alice's side, giving her $\sigma_{BA'} = U_{A\rightarrow BA'}\rho_A U^\dagger_{A\rightarrow BA'}$, and in a second step use State Splitting to do an optimal state transfer of the B-part to Bob's side, such that he holds $\sigma_B = \mathcal{E}_{A\rightarrow B}(\rho_A)$ in the end. This simulates the channel $\mathcal{E}_{A\rightarrow B}$. To prove the Quantum Reverse Shannon Theorem, it is then sufficient to show that the classical communication rate of the State Splitting protocol is $C_E(\mathcal{E})$.

---

[2] More precisely, State Merging corresponds to the task of obtaining the state $\rho_{B'BR} = (\text{id}_{A\rightarrow B'} \otimes \text{id}_{BR})\rho_{ABR}$, where ρ_{ABR} is a purification of ρ_{AB}, R is a reference, and BB' is held by Bob.

[3] More precisely, State Splitting corresponds to the task of obtaining the state $\rho_{ABR} = (\text{id}_{B'\rightarrow A} \otimes \text{id}_{BR})\rho_{B'BR}$, where $\rho_{B'BR}$ is a purification of $\rho_{B'B}$, R is a reference, A is held by Alice and B is held by Bob.

We realize this idea in two steps. Firstly, we propose a new version of State Merging/Splitting (since the known protocols are not good enough to achieve a classical communication rate of C_E). For the analysis we require a 'Decoupling Theorem', which is optimal in the most general case, the so called one-shot case [4, 6, 10]. To quantify the resources needed for one-shot State Merging/Splitting, we make use of the 'Smooth Rényi Entropy Calculus' [8, 16, 19, 27, 28]. Secondly, we use the Post-Selection Technique to show that our protocol for one-shot State Splitting for a particular de Finetti type input state is sufficient to asymptotically simulate the channel $\mathcal{E}$ for a classical communication rate of C_E for any input. This then completes the proof of the Quantum Reverse Shannon Theorem.

This extended abstract is structured as follows. In Section 2 we introduce our notation and give some definitions. In particular, we review the smooth entropy measures that we need in the following. Our results about one-shot State Splitting are then discussed in Section 3. Finally, we sketch our proof of the Quantum Reverse Shannon Theorem in Section 4 (journal reference [5]).

2   (Smooth) Entropy Measures – Notation and Definitions

We assume that all Hilbert spaces, in the following denoted $\mathcal{H}$, are finite-dimensional. We write $|A|$ for the dimension of $\mathcal{H}_A$. The the set of linear, non-negative operators on $\mathcal{H}$ is denoted by $\mathcal{P}(\mathcal{H})$. $\mathbb{1}$ denotes the identity in $\mathcal{P}(\mathcal{H})$. We define the sets of normalized states $\mathcal{S}_=(\mathcal{H}) = \{\rho \in \mathcal{P}(\mathcal{H}) : \mathrm{tr}\rho = 1\}$ and subnormalized states $\mathcal{S}_\leq(\mathcal{H}) = \{\rho \in \mathcal{P}(\mathcal{H}) : \mathrm{tr}\rho \leq 1\}$).

In quantum information theory one usually makes the assumption that the resources are independent and identically distributed (i.i.d.) and is interested in asymptotic rates. In this case many operational quantities can be expressed in terms of a few information measures (which are usually based on the von Neumann entropy). In order to overcome the asymptotic and i.i.d. assumption, the Smooth Rényi Entropy Calculus was introduced by Renner et al. [19, 20, 21].

Recall the following standard definitions. The *von Neumann entropy* of $\rho \in \mathcal{S}_=(\mathcal{H})$ is defined as $H(\rho) = -\mathrm{tr}(\rho \log \rho)$. The *quantum relative entropy* of $\rho \in \mathcal{S}_\leq(\mathcal{H})$ with respect to $\sigma \in \mathcal{P}(\mathcal{H})$ is given by $D(\rho\|\sigma) = \mathrm{tr}(\rho \log \rho) - \mathrm{tr}(\rho \log \sigma)$. The *conditional von Neumann entropy* of A given B for $\rho_{AB} \in \mathcal{S}_=(\mathcal{H})$ is defined as $H(A|B)_\rho = -D(\rho_{AB}\|\mathbb{1}_A \otimes \rho_B)$. The *mutual information* between A and B for $\rho_{AB} \in \mathcal{S}_=(\mathcal{H})$ is given by $I(A:B)_\rho = D(\rho_{AB}\|\rho_A \otimes \rho_B)$. Note that we can also write

$$H(A|B)_\rho = -\inf_{\sigma_B} D(\rho_{AB}\|\mathbb{1}_A \otimes \sigma_B) \tag{3}$$

$$I(A:B)_\rho = \inf_{\sigma_B} D(\rho_{AB}\|\rho_A \otimes \sigma_B) , \tag{4}$$

where $\sigma_B \in \mathcal{S}_=(\mathcal{H}_B)$.

Following Datta [8] we define the *max-relative entropy* of $\rho \in \mathcal{S}_\leq(\mathcal{H})$ with respect to $\sigma \in \mathcal{P}(\mathcal{H})$ as

$$D_{\max}(\rho\|\sigma) = \inf\{\lambda \in \mathbb{R} : 2^\lambda \cdot \sigma \geq \rho\} . \tag{5}$$

The *conditional min-entropy* [19] of A given B for $\rho_{AB} \in \mathcal{S}_\leq(\mathcal{H}_{AB})$ is defined as

$$H_{\min}(A|B)_\rho = -\inf_{\sigma_B} D_{\max}(\rho_{AB}\|\mathbb{1}_A \otimes \sigma_B) = \sup_{\sigma_B} H_{\min}(A|B)_{\rho|\sigma} , \qquad (6)$$

where $H_{\min}(A|B)_{\rho|\sigma} = -D_{\max}(\rho_{AB}\|\mathbb{1}_A \otimes \sigma_B)$ and $\sigma_B \in \mathcal{S}_=(\mathcal{H}_B)$. In the special case where B is trivial, we have $H_{\min}(A)_\rho = -\log \|\rho_A\|_\infty$.[4] The *max-information* that B has about A for $\rho_{AB} \in \mathcal{S}_\leq(\mathcal{H}_{AB})$ is defined as

$$I_{\max}(A:B)_\rho = \inf_{\sigma_B} D_{\max}(\rho_{AB}\|\rho_A \otimes \sigma_B) , \qquad (7)$$

where $\sigma_B \in \mathcal{S}_=(\mathcal{H}_B)$. Note that unlike the mutual information, this definition is not symmetric. The smooth entropy measures are defined by extremizing the non-smooth measures over a set of nearby states, where our notion of nearby is expressed in terms of the *purified distance*. For $\rho, \sigma \in \mathcal{S}_\leq(\mathcal{H})$ it is defined as [28]

$$P(\rho,\sigma) = \sqrt{1 - \bar{F}(\rho,\sigma)^2} , \qquad (8)$$

where $\bar{F}(\cdot,\cdot)$ denotes the *generalized fidelity* (which equals the standard fidelity if at least one of the states is normalized),

$$\bar{F}(\rho,\sigma) = \left\| \sqrt{\rho \oplus (1 - \mathrm{tr}\rho)} \sqrt{\sigma \oplus (1 - \mathrm{tr}\sigma)} \right\|_1 = \left\| \sqrt{\rho}\sqrt{\sigma} \right\|_1 + \sqrt{(1 - \mathrm{tr}\rho)(1 - \mathrm{tr}\sigma)} . \qquad (9)$$

The purified distance is a distance measure. Henceforth we call $\rho, \sigma \in \mathcal{S}_\leq(\mathcal{H})$ ϵ-close if $P(\rho,\sigma) \leq \epsilon$ and denote this by $\rho \approx_\epsilon \sigma$. Miscellaneous properties of the purified distance are stated in Appendix A of [5]. We use the purified distance to specify a ball of subnormalized density operators around $\rho \in \mathcal{S}_\leq(\mathcal{H})$:

$$\mathcal{B}^\epsilon(\rho) = \{\bar{\rho} \in \mathcal{S}_\leq(\mathcal{H}) : P(\rho,\bar{\rho}) \leq \epsilon\} .$$

For any $\epsilon \geq 0$, the *smooth conditional min-entropy* [19] of A given B for $\rho_{AB} \in \mathcal{S}_\leq(\mathcal{H}_{AB})$ is defined as

$$H_{\min}^\epsilon(A|B)_\rho = \sup_{\bar{\rho}_{AB}} H_{\min}(A|B)_{\bar{\rho}} , \qquad (10)$$

where $\bar{\rho}_{AB} \in \mathcal{B}^\epsilon(\rho_{AB})$. The *smooth max-information* that B has about A for $\rho_{AB} \in \mathcal{S}_\leq(\mathcal{H}_{AB})$ is defined as

$$I_{\max}^\epsilon(A:B)_\rho = \inf_{\bar{\rho}_{AB}} I_{\max}(A:B)_{\bar{\rho}} , \qquad (11)$$

where $\bar{\rho}_{AB} \in \mathcal{B}^\epsilon(\rho_{AB})$. These smooth entropy measures can be seen as generalizations of their corresponding von Neumann quantities in the sense of Lemmata B.17 and B.22 in [5], i.e. they asymptotically converge to the conditional von Neumann entropy and the mutual information resp. In Section 3 we give an operational meaning to the smooth max-information (see Theorem 3.2 and Theorem 3.3).[5]

---

[4] $\|\rho_A\|_\infty$ denotes the maximal eigenvalue of ρ_A.
[5] For an operational meaning of the smooth conditional min-entropy see e.g. [4, 19].

3   One-Shot Quantum State Splitting

State Merging, State Splitting and other related quantum information processing primitives are discussed in detail in [1, 13, 14, 18]. In contrast to the existing literature, we are not only interested in asymptotic rates, but in a (tight) one-shot protocol for State Splitting. For more details see [5].

Definition 3.1 (Quantum State Splitting with embezzling states). *Consider a system consisting of three parties; Alice, Bob and a reference. Let $\rho_{ACR} \in \mathcal{S}_\le(\mathcal{H}_{ACR})$ be pure, where the reference is denoted by R and Alice has AC. Let B be an ancilla at Bob's side of the same size as C. Furthermore let Alice and Bob share an embezzling state of index k that lives on an additional register $\overline{AB}$. A process is called* embezzling State Splitting *of ρ_{ACR} with error ϵ if it consists of applying local operations at Alice's side, local operations at Bob's side, sending q qubits from Alice to Bob and outputs a state $\omega_{ABR} \approx_\epsilon \rho_{ABR}$ with $\rho_{ABR} = (\mathrm{id}_{C \to B} \otimes \mathrm{id}_{AR})\rho_{ACR}$. q is called* quantum communication cost *and $e_b = \lceil \log k \rceil$ is the* embezzling cost.

Theorem 3.2. *Let $\rho_{ACR} = |\psi\rangle\langle\psi|_{ACR} \in \mathcal{S}_\le(\mathcal{H}_{ACR})$, $\epsilon > 0$, $\epsilon' \ge 0$ and $\epsilon'' > 0$. Then there exists an embezzling State Splitting protocol of ρ_{ACR} with a quantum communication cost of*

$$q \le \frac{1}{2} I^{\epsilon'}_{\max}(C : R)_\rho + \log \frac{1}{\epsilon} + 4 + \log \log |C| \tag{12}$$

and an embezzling cost of

$$e_b \ge \lceil (\log |C| - \log \frac{1}{\epsilon})^{1/\epsilon''} \rceil \tag{13}$$

for an error of at most $\epsilon' + \sqrt{\epsilon} + |C|^{-1/2} + \sqrt{\epsilon''}$.

For a proof see Theorem III.2 in [5]. The following theorem shows that this is optimal up to small additive terms.

Theorem 3.3 (Converse). *There does not exist a State Splitting protocol with error ϵ and a quantum communication cost smaller than*

$$q = \frac{1}{2} I^{\epsilon}_{\max}(C : R)_\rho . \tag{14}$$

For a proof see Theorem III.9 in [5].

4   The Quantum Reverse Shannon Theorem

We now present our main result; a proof of the Quantum Reverse Shannon Theorem. Let $\mathcal{E}_{A \to B}$ be a quantum channel with

$$\mathcal{E}_{A \to B} : \mathcal{S}(\mathcal{H}_A) \to \mathcal{S}(\mathcal{H}_B)$$
$$\rho_A \mapsto \mathcal{E}_{A \to B}(\rho_A) ,$$

where we want to think of A being at Alice's side and B being at Bob's side. The Quantum Reverse Shannon Theorem states that if we have an embezzling state (of arbitrary large index) between Alice and Bob, we can asymptotically simulate $\mathcal{E}_{A\to B}$ only using local operations at Alice's side, local operations at Bob's side, and a classical communication rate (from Alice to Bob) of

$$C_E(\mathcal{E}) = \max_\rho I(B:R)_{(\mathcal{E}\otimes\mathrm{id})(\Phi)} , \qquad (15)$$

where Φ_{AR} is a purification of ρ_A.[6] Using Stinespring's dilation [26], we can think of $\mathcal{E}_{A\to B}$ as

$$\mathcal{E}_{A\to B}(\rho_A) = \mathrm{tr}_{B'}(U_{A\to BA'}\rho_A U^\dagger_{A\to BA'}) ,$$

where A' is an additional register with $|A'| \le |A||B|$ and $U_{A\to BA'}$ is some isometry from A to BA'. The idea of our proof is to first simulate the quantum channel locally at Alice's side, giving us $\sigma_{BA'} = U_{A\to BA'}\rho_A U^\dagger_{A\to BA'}$, and then use embezzling State Splitting to do an optimal state transfer of the B-part to Bob's side, such that he holds $\sigma_B = \mathcal{E}_{A\to B}(\rho_A)$ in the end. Note that we can replace the quantum communication in the embezzling State Splitting protocol by twice as much classical communication since we have free entanglement and can therefore use teleportation. Although the free entanglement is given in the form of embezzling states, maximally entangled states can be created without any (additional) communication (Proposition 1.2). Because the quantum reverse Shannon theorem makes an asymptotic statement, we have to make our considerations for a general $n \in \mathbb{N}$. The idea is then to show the existence of a protocol $\mathcal{F}^n_{A\to B}$ that is arbitrarily close to $\mathcal{E}^{\otimes n}_{A\to B}$ for $n \to \infty$, has a classical communication rate of $C_E(\mathcal{E})$ and works for any input. We do this by using the post-selection technique (Proposition D.4 in [5]).

Since the post-selection technique only applies to permutation invariant maps, we need to ensure that the protocol that we want to use for the simulation of $\mathcal{E}^{\otimes n}_{A\to B}$, can be made to act symmetrically on the n-partite input system $\mathcal{H}^{\otimes n}_A$. This can be done by inserting a symmetrization step, that uses maximally entangled states (cf. Figure 1). For more details see Section IV of [5].

Let $\delta > 0$. The way to go is to show the existence of a permutation invariant $\mathcal{F}^n_{A\to B}$ with the desired properties as discussed above (i.e. it should be local and have a classical communication rate of $C_E(\mathcal{E})$) and

$$\|((\mathcal{E}^{\otimes n}_{A\to B} - \mathcal{F}^n_{A\to B}) \otimes \mathrm{id}_{RR'})(\zeta^n_{ARR'})\|_1 \le \delta(n+1)^{-(|A|^2-1)} , \qquad (16)$$

where $\zeta^n_{ARR'}$ is a purification of $\zeta^n_{AR} = \int \rho^{\otimes n}_{AR} d(\rho_{AR})$, ρ_{AR} is pure and $d(.)$ is the measure on the normalized pure states on $\mathcal{H}_{AR}$ induced by the Haar measure on the unitary group acting on $\mathcal{H}_{AR}$, normalized to $\int d(.) = 1$. Once we have achieved this it follows from the post-selection technique (Proposition D.4

---

[6] Since all purifications give the same amount of entropy, we do not need to specify which one we use.

138   M. Berta, M. Christandl, and R. Renner

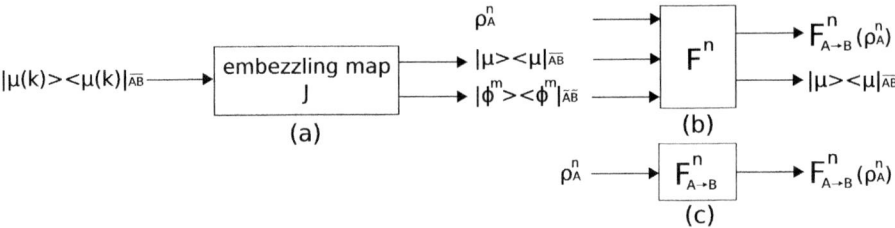

Fig. 1. (a) $\mathcal{J}$ is the map that embezzles m maximally entangled states $|\Phi^m\rangle\langle\Phi^m|_{\tilde{A}\tilde{B}}$ out of $|\mu\rangle\langle\mu|_{\overline{AB}}$. These maximally entangled states are then used for a symmetrization step. (b) The whole protocol $\mathcal{F}^n_{A\to B}$ (that should simulate $\mathcal{E}^{\otimes n}_{A\to B}$) a priori takes $\rho^n_A \otimes |\mu\rangle\langle\mu|_{\overline{AB}} \otimes |\Phi^m\rangle\langle\Phi^m|_{\tilde{A}\tilde{B}}$ as an input. But since this input is constant on all registers except for A, we can think of the map as in (c), namely as a CPTP map $\mathcal{F}^n_{A\to B}$ which only takes the input ρ^n_A.

in [5]) that $\|\mathcal{E}^{\otimes n}_{A\to B} - \mathcal{F}^n_{A\to B}\|_\diamond \le \delta$, and this is what we want to show.[7] Since we only need to consider the particular de Finetti type input state $\zeta^n_{ARR'}$ to show (16), we can use the State Splitting protocol of Theorem 3.2 for $\zeta^n_{ARR'}$ to construct $\mathcal{F}^n_{A\to B}$. A lengthy calculation (see Section IV of [5]) shows, that a classical communication rate of $C_E(\mathcal{E})$ is sufficient to get (16). This concludes the proof of the Quantum Reverse Shannon Theorem.

Acknowledgments

We thank Jürg Wullschleger and Andreas Winter for inspiring discussions. MB and MC are supported by the Swiss National Science Foundation (grant PP00P2-128455) and the German Science Foundation (grants CH 843/1-1 and CH 843/2-1). RR acknowledges support from the Swiss National Science Foundation (grant 200021-119868).

References

[1] Abeyesinghe, A., Devetak, I., Hayden, P., Winter, A.: The mother of all protocols: Restructuring quantum information's family tree. Proc. R. Soc. A 465(2108), 2537 (2009), arXiv.org:quant-ph/0606225
[2] Bennett, C.H., Devetak, I., Harrow, A.W., Shor, P.W., Winter, A.: The quantum reverse Shannon theorem (2006), arXiv.org:quant-ph/0912.5537
[3] Bennett, C.H., Shor, P.W., Smolin, J.A., Thapliyal, A.V.: Entanglement-assisted capacity of a quantum channel and the reverse Shannon theorem. IEEE Trans. Inf. Theory 48(10), 2637 (2002), arXiv.org:quant-ph/0106052
[4] Berta, M.: Single-shot quantum state merging, Diploma thesis ETH Zurich (2008), arXiv.org:quant-ph/0912.4495

[7] For a precise definition of the Diamond norm see [15].

[5] Berta, M., Christandl, M., Renner, R.: A Conceptually Simple Proof of the Quantum Reverse Shannon Theorem (2009), arXiv.org:quant-ph/0912.3805, submitted to Comm. Math. Phys. (2009)

[6] Berta, M., Dupuis, F., Renner, R., Wullschleger, J.: Optimal decoupling (2010) (in preparation)

[7] Christandl, M., König, R., Renner, R.: Post-selection technique for quantum channels with applications to quantum cryptography. Phys. Rev. Lett 102, 20504 (2009), arXiv.org:quant-ph/0809.3019

[8] Datta, N.: Min- and max- relative entropies and a new entanglement monotone. IEEE Trans. Inf. Theory 55(6), 2816 (2009), arXiv.org:quant-ph/0803.2770

[9] Devetak, I.: The private classical capacity and quantum capacity of a quantum channel. IEEE Trans. Inf. Theory 51, 44 (2005), arXiv:quant-ph/0304127

[10] Dupuis, F.: The Decoupling Approach to Quantum Information Theory. PhD thesis, Université de Montréal (2009), arXiv.org:quant-ph/1004.1641

[11] Harrow, A.W.: Entanglement spread and clean resource inequalities. Proc. XVI Int. Cong. Math. Phys. 536 (2009), arXiv.org:quant-ph/0909.1557

[12] Holevo, A.S.: The capacity of the quantum communication channel with general signal states. IEEE Trans. Inf. Theory 44, 269 (1998), arXiv.org:quant-ph/9611023

[13] Horodecki, M., Oppenheim, J., Winter, A.: Partial quantum information. Nature 436, 673–676 (2005), arXiv.org:quant-ph/0505062

[14] Horodecki, M., Oppenheim, J., Winter, A.: Quantum state merging and negative information. Comm. Math. Phys 269, 107 (2006), arXiv.org:quant-ph/0512247

[15] Kitaev, A.: Quantum computations: algorithms and error correction. Russian Math. Surveys 52, 1191 (1997)

[16] König, R., Renner, R., Schaffner, C.: The operational meaning of min- and max-entropy. IEEE Trans. Inf. Theory 55(9), 4337 (2009), arXiv.org:quant-ph/0807.1338

[17] Lloyd, S.: Capacity of the noisy quantum channel. Phys. Rev. A 55, 1613 (1997), arXiv.org:quant-ph/9604015

[18] Oppenheim, J.: State redistribution as merging: introducing the coherent relay (2008), arXiv.org:quant-ph/0805.1065

[19] Renner, R.: Security of Quantum Key Distribution. PhD thesis, ETH Zurich (2005), arXiv.org:quant-ph/0512258

[20] Renner, R.S., König, R.: Universally Composable Privacy Amplification Against Quantum Adversaries. In: Kilian, J. (ed.) TCC 2005. LNCS, vol. 3378, pp. 407–425. Springer, Heidelberg (2005)

[21] Renner, R., Wolf, S.: Smooth Rényi entropy and applications. In: Proc. IEEE Int. Symp. Inf. Theory, vol. 233 (2004)

[22] Schumacher, B., Westmoreland, M.D.: Sending classical information via noisy quantum channels. Phys. Rev. A 56, 131 (1997)

[23] Shannon, C.E.: A mathematical theory of communication. Bell. Syst. Tech. J. 423, 379–423, 623–656 (1948)

[24] Shor, P.W.: The quantum channel capacity and coherent information. In: Lecture notes, MSRI Workshop on Quantum Computation (2002)

[25] Slepian, D., Wolf, J.: Noiseless coding of correlated information sources. IEEE Trans. Inf. Theory 19, 461 (1971)

[26] Stinespring, W.: Positive function on C*-algebras. Proc. Amer. Math. Soc. 6, 211 (1955)

[27] Tomamichel, M., Colbeck, R., Renner, R.: A fully quantum asymptotic equipartition property. IEEE Trans. Inf. Theory 55(12), 5840 (2009), arXiv.org:quant-ph/0811.1221
[28] Tomamichel, M., Colbeck, R., Renner, R.: Duality between smooth min- and max-entropies. IEEE Trans. Inf. Theory 56(9), 4674 (2010), arXiv.org:quant-ph/0907.5238
[29] van Dam, W., Hayden, P.: Universal entanglement transformations without communication. Phys. Rev. A, Rapid Comm. 67, 060302(R) (2003), arXiv.org:quant-ph/0201041

Geometric Entanglement of Symmetric States and the Majorana Representation

Martin Aulbach[1,2], Damian Markham[3], and Mio Murao[4,5]

[1] The School of Physics and Astronomy, University of Leeds, Leeds LS2 9JT, United Kingdom
pyma@leeds.ac.uk
[2] Department of Physics, University of Oxford, Clarendon Laboratory, Oxford OX1 3PU, United Kingdom
m.aulbach1@physics.ox.ac.uk
[3] CNRS, LTCI, Telecom ParisTech, 23 Avenue d'Italie, 75013 Paris, France
markham@telecom-paristech.fr
[4] Department of Physics, Graduate School of Science, The University of Tokyo, Tokyo 113-0033, Japan
murao@phys.s.u-tokyo.ac.jp
[5] Institute for Nano Quantum Information Electronics, The University of Tokyo, Tokyo 113-0033, Japan

Abstract. Permutation-symmetric quantum states appear in a variety of physical situations, and they have been proposed for quantum information tasks. This article builds upon the results of [New J. Phys. 12 (2010) 073025], where the maximally entangled symmetric states of up to twelve qubits were explored, and their amount of geometric entanglement determined by numeric and analytic means. For this the Majorana representation, a generalization of the Bloch sphere representation, can be employed to represent symmetric n qubit states by n points on the surface of a unit sphere. Symmetries of this point distribution simplify the determination of the entanglement, and enable the study of quantum states in novel ways. Here it is shown that the duality relationship of Platonic solids has a counterpart in the Majorana representation, and that in general maximally entangled symmetric states neither correspond to anticoherent spin states nor to spherical designs. The usability of symmetric states as resources for measurement-based quantum computing is also discussed.

Keywords: Majorana representation, geometric measure, symmetric, entanglement, anticoherent, spherical design.

1 Introduction

Multipartite entanglement is a crucial resource for many tasks in quantum information science, but its quantification is difficult due to the existence of different types of entanglement [17]. It is therefore unsurprising that many different entanglement measures have been proposed in order to quantify the amount of

W. van Dam et al. (Eds.): TQC 2010, LNCS 6519, pp. 141–158, 2011.
© Springer-Verlag Berlin Heidelberg 2011

entanglement of multipartite quantum states [25]. Here we build upon our results about highly and maximally entangled permutation-symmetric quantum states in terms of the geometric measure of entanglement [4]. This restriction to a subset of quantum states – studied under a particular entanglement measure – makes it possible to gain strong results [4,41], and to find a rare visual representation of multipartite entanglement.

Permutation-symmetric quantum states are invariant under any permutation of their subsystems. Such states appear in many-body physics, and they have found use in leader election [13]. Furthermore, they have been actively implemented experimentally [50,67], and their symmetric properties facilitate the analysis of entanglement [8,24,26,38,43,59]. In order to analyze the usefulness of symmetric states for measurement-based quantum computation (MBQC) [46], the geometric measure of entanglement is particularly suited, because the classification of states as MBQC-resources has been performed in terms of this measure [20,44,45].

The central tool for our analysis of symmetric entanglement is the Majorana representation [36], a generalization of the Bloch sphere representation of single qubits. By means of this representation any n qubit symmetric state can be unambiguously mapped to n points on the surface of the unit sphere. Recently the Majorana representation has been used to search for and characterize different classes of SLOCC entanglement [3,8,38,43], which is related to the classification of phases in spinor condensates [6,38]. It has also been employed to search for the "least classical" state of a spin-j system [19], and the solutions of this problem are intimately related to the maximally entangled symmetric states. Furthermore, the Majorana representation has been used for the study of spherical designs [11], Berry phases in high spin systems [21], quantum chaos [22,34], optimal resources for reference frame alignment [29], phase estimation [28], phases in spinor BEC [6,37], classicality in terms of the discriminability of states [40], for finding solutions to the Lipkin-Meshkov-Glick model [53] and for finding efficient proofs of the Kochen-Specker theorem [70].

The article is organized as follows: In Sect. 2 we briefly recapitulate the geometric measure of entanglement. This is followed by Sect. 3 where the geometric entanglement of permutation-symmetric states and its implications for MBQC is discussed. In Sect. 4 the Majorana representation is introduced for symmetric states of n qubits, which is followed by Sect. 5 which reviews our analytical and numerical findings that we recently published in [4]. In Sect. 6 the usefulness of the Majorana representation is demonstrated for highly entangled symmetric states whose point distributions are described by Platonic solids. The entanglement of such states is particularly easy to determine with the known theoretical results, and it is found that there exists an intriguing analogy with the dual polyhedra of the Platonic solids [64]. Anticoherent spin states [69] and the mathematical concept of spherical designs [11] are briefly mentioned, and it is shown that in general the maximally entangled symmetric states do not represent anticoherent states or spherical designs. Finally, Sect. 7 concludes this article with a summary of our results.

2 Geometric Measure of Entanglement

The geometric measure of entanglement is a distance-like entanglement measure in the sense that it assesses the entanglement in terms of the remoteness from the set of separable states [60]. It is defined as the maximal overlap of a normalized pure state with all normalized pure product states [7,54,63].

$$E_g(|\psi\rangle) = \min_{|\lambda\rangle \in \mathcal{H}_{\mathrm{SEP}}} -\log_2 |\langle\lambda|\psi\rangle|^2 \ . \tag{1}$$

A product state closest to $|\psi\rangle$ is denoted by $|\Lambda_\psi\rangle$, and it should be kept in mind that a given $|\psi\rangle$ can have more than one closest product state. The problem of maximizing the entanglement can be written as a max-min-problem:

$$\begin{aligned} E_g^{\mathrm{max}} &= \max_{|\psi\rangle \in \mathcal{H}} \min_{|\lambda\rangle \in \mathcal{H}_{\mathrm{SEP}}} -\log_2 |\langle\lambda|\psi\rangle|^2 \\ &= \max_{|\psi\rangle \in \mathcal{H}} -\log_2 |\langle\Lambda_\psi|\psi\rangle|^2 = -\log_2 |\langle\Lambda_\Psi|\Psi\rangle|^2 \ . \end{aligned} \tag{2}$$

The geometric measure is closely related to the robustness of entanglement R [61] and the relative entropy of entanglement E_R [60], two other distance-like entanglement measures. The inequalities $E_g \leq E_R \leq \log_2(1 + R)$ hold for all states [9,24,62], and they become equalities for stabilizer states, Dicke states and permutation-antisymmetric basis states [23,24,39]. Some advantages of the geometric measure are its comparatively easy calculation, its applications in related fields of physics [33,47,55], and its operational interpretations, e.g. in local state discrimination [23], additivity of channel capacities [65] and for the classification of states as resources for measurement-based quantum computation (MBQC)[20,44,45].

A general quantum state of a finite-dimensional system can be cast as $|\psi\rangle = \sum_i a_i |i\rangle$ with complex coefficients a_i and an orthonormal basis $\{|i\rangle\}$. The state $|\psi\rangle$ is called real if (for a given basis) the a_i are all real, and positive if the a_i are all positive. Every positive state $|\psi\rangle$ has at least one positive closest product state $|\Lambda_\psi\rangle$ [4,68], a result which simplifies the determination of their entanglement.

3 Permutation-Symmetric States

Permutation-symmetric quantum states are states that are invariant under any permutation of their subsystems, i.e. $P|\psi\rangle = |\psi\rangle$ for all $P \in S_N$. For n qubits the Hilbert space of symmetric states is spanned by the Dicke states, the equally weighted sums of all permutations of computational basis states with $n-k$ qubits being $|0\rangle$ and k being $|1\rangle$ [14,58].

$$|S_{n,k}\rangle = \binom{n}{k}^{-1/2} \sum_{\mathrm{perm}} \underbrace{|0\rangle|0\rangle \cdots |0\rangle}_{n-k} \underbrace{|1\rangle|1\rangle \cdots |1\rangle}_{k} \ , \tag{3}$$

with $0 \leq k \leq n$. A general pure symmetric state of n qubits is a linear combination of the $n + 1$ symmetric basis states $|S_{n,k}\rangle$. We will abbreviate this notation to $|S_k\rangle$ whenever the number of qubits is clear.

It was recently found that all closest product states of multipartite (≥ 3 parts) symmetric states are symmetric themselves, and that bipartite symmetric states have at least one symmetric closest product state [26]. Furthermore, it can be shown that positive symmetric states have at least one positive symmetric closest product state [24]. These results considerably reduce the complexity of finding the closest product state and thus the entanglement of a symmetric state.

The theoretical and experimental analysis of symmetric state entanglement, e.g. as entanglement witnesses or in experimental setups [30,31,50,67], is valuable, because symmetric states appear in many-body physics. For example, the ground state of the Lipkin-Meshkov-Glick model is permutation-invariant, and its entanglement has been quantified in term of the geometric measure [48].

3.1   Bounds on Maximal Entanglement

In this subsection we will briefly discuss the known upper and lower bounds on the maximal possible amount of geometric entanglement. It should however be kept in mind that the maximally entangled state and its amount of entanglement depends on the chosen entanglement measure [49].

The maximal possible entanglement of general n qubit states scales linearly with the number of qubits, namely

$$\tfrac{n}{2} \leq E_{\mathrm{g}}^{\mathrm{max}} \leq n - 1 \ . \tag{4}$$

The left-hand side of the inequality is clear from the trivial example of an n qubit state (n even) composed of $\frac{n}{2}$ bipartite Bell states, or from 2D cluster states [39]. The upper bound was derived in [27]. It is also known that most n qubit states are much closer to the upper bound than to the lower bound. More precisely, for $n > 10$ qubits the overwhelming majority of states have entanglement $E_{\mathrm{g}} > n - 2\log_2(n) - 3$ [20].

For symmetric states a trivial lower bound can be derived from the Dicke states. A closest product state of $|S_{n,k}\rangle$ is known [24] to be

$$|\Lambda\rangle = \left(\sqrt{\tfrac{n-k}{n}} \, |0\rangle + \sqrt{\tfrac{k}{n}} \, |1\rangle \right)^{\otimes n} \ . \tag{5}$$

From this the entanglement follows as

$$E_{\mathrm{g}}(|S_{n,k}\rangle) = \log_2 \left(\frac{\left(\tfrac{n}{k}\right)^k \left(\tfrac{n}{n-k}\right)^{n-k}}{\binom{n}{k}} \right) \ . \tag{6}$$

The maximally entangled Dicke state is $|S_{n,n/2}\rangle$ for even n and the two equivalent states $|S_{n,\lfloor n/2 \rfloor}\rangle$ and $|S_{n,\lceil n/2 \rceil}\rangle$ for odd n. Their Stirling approximation for large n yields $E_{\mathrm{g}}^{\mathrm{max}} \geq \log_2 \sqrt{n\pi/2}$. An upper bound to the geometric measure of symmetric n qubit states has been derived from the decomposition of the identity on symmetric subspace, yielding $E_{\mathrm{g}}^{\mathrm{max}} \leq \log_2(n+1)$, see e.g. [52]. An alternative proof with the benefit of being visually accessible by means of the Majorana representation will be given in Theorem 2.

Combining these bounds, it is seen that the maximal symmetric entanglement of n qubits scales as

$$\log_2 \sqrt{\tfrac{n\pi}{2}} \le E_{\mathrm{g}}^{\max} \le \log_2(n+1) \ , \tag{7}$$

i.e. polylogarithmically between $\mathcal{O}(\log \sqrt{n})$ and $\mathcal{O}(\log n)$. Numerical evidence suggests that the actual values are much closer to the upper bound than to the lower bound, and $E_{\mathrm{g}}^{\max} \gtrsim \log_2(n+1) - 0.775$ can be considered a reliable lower bound [41].

3.2   Resources for MBQC

We have seen that the maximal entanglement of symmetric states scales much slower than that of general states, namely logarithmically rather than linearly. This need not be a disadvantage for symmetric states, though, and in fact could render them useful for MBQC [46], because it was shown that if the entanglement of a state is too large, then it cannot be a good resource for MBQC. More specifically, if the n qubit entanglement scales larger than $n-\delta$ for some constant δ, then such a computation can be simulated efficiently classically [20]. This rules out many general quantum states as MBQC resources, but not symmetric ones.

On the other hand, universal resources for MBQC must be maximally entangled in a certain sense [44,45]. Considering the qualitative departure of the scaling relation (7) from (4), it is questionable whether symmetric states are sufficiently entangled to be MBQC resources. Indeed, permutation-symmetric states can be ruled out as exact, deterministic MBQC resources, because their entanglement does not scale faster-than-logarithmically [4,45]. Somewhat weaker requirements are imposed upon approximate, stochastic MBQC resources [44], although this generally leads only to a small extension of the class of suitable resources in the vicinity of exact, deterministic resources (e.g. 2D cluster states with holes). It is therefore believed that symmetric states cannot be used even for approximate, stochastic MBQC.

As an example, we will show that Dicke states with a fixed number of excitations cannot be useful for ϵ-approximate, deterministic MBQC [44]. Roughly speaking, ϵ-approximate universal resource states can be converted into any other state by LOCC with an inaccuracy of at most ϵ. The ϵ-version of the geometric measure[1] is defined as [44]

$$E_{\mathrm{G}}^{\epsilon}(\rho) = \min\{E_{\mathrm{G}}(\sigma) \,|\, D(\rho, \sigma) \le \epsilon\} \ , \tag{8}$$

where D is a distance that is "strictly related to the fidelity", meaning that for any two states ρ and σ, $D(\rho, \sigma) \le \epsilon \Rightarrow F(\rho, \sigma) \ge 1 - \eta(\epsilon)$, where $0 \le \eta(\epsilon) \le 1$ is a strictly monotonically increasing function with $\eta(0) = 0$. $E_{\mathrm{G}}^{\epsilon}(\rho)$ can be

---

[1] There exist two common definitions of the geometric measure, which we distinguish by denoting $E_{\mathrm{g}}(|\psi\rangle) = -\log_2 |\langle \Lambda_\psi | \psi \rangle|^2$ and $E_{\mathrm{G}}(|\psi\rangle) = 1 - |\langle \Lambda_\psi | \psi \rangle|^2$. The former is used throughout this article, with the only exception being this subsection, where E_{G} is used in order to be consistent with the notation of [44].

understood as the guaranteed entanglement obtained from a preparation of ρ with inaccuracy ϵ. One possible choice of D is the trace distance, which for pure states reads $D_t(|\psi\rangle, |\phi\rangle) = \sqrt{1 - |\langle\psi|\phi\rangle|^2} = \sqrt{1 - F}$, where F is the fidelity. In this case one can choose $\eta(\epsilon) = \epsilon^2$.

As shown in Example 1 of [44], the family of W states $\Psi_W = \{|W_n\rangle\}_n$, with $|W_n\rangle \equiv |S_{n,1}\rangle$, is not an ϵ-approximate universal resource for $\eta(\epsilon) \lesssim 0.001$. This result can be generalized to all families of Dicke states $\Psi_{S_k} = \{|S_{n,k}\rangle\}_n$ with a fixed number of excitations k.

Example 1. For any fixed $k \in \mathbb{N}$ the family of Dicke states $\Psi_{S_k} = \{|S_{n,k}\rangle\}_n$ cannot be an ϵ-approximate universal MBQC resource for $\eta(\epsilon) \lesssim 0.001 \, k^{-3/2}$.

Proof. Using (6) and the Stirling approximation for high n, the asymptotic geometric entanglement of the family Ψ_{S_k} is found to be

$$E_G(\Psi_{S_k}) = 1 - \frac{k^k}{e^k k!} \quad . \tag{9}$$

Specifically, the amount of geometric entanglement remains finite for arbitrary values of n, allowing us to apply Proposition 3 and Theorem 1 of [44] to show that the necessary condition for ϵ-approximate deterministic universality,

$$E_G(\Psi_{S_k}) > 1 - 4\eta^{1/3} + 3.4\eta^{2/3} \quad , \tag{10}$$

is violated for $\eta(\epsilon) \lesssim 0.001 \, k^{-3/2}$. $\qquad\square$

Of course, it should be noted that many other quantum information tasks are not restricted by the requirements of MBQC-universality, and that highly entangled symmetric states can therefore be valuable resources for such tasks.

4   Majorana Representation of Symmetric States

The classical angular momentum **J** of a physical system can be represented by a single point on the surface of the unit sphere in $\mathbb{R}^3$, corresponding to the direction of **J**. Quantum mechanics does not allow for such a simple representation, but it is possible to uniquely represent a pure state of spin-j by $2j$ undistinguishable points on the sphere [36]. This is a generalization of the Bloch sphere representation of a qubit. An equivalent representation can be shown to exist for symmetric states of n spin-$(1/2)$ particles [5,36], with an isomorphism mediating between all states of a spin-j particle and the symmetric states of $2j$ qubits.

Hence, this "Majorana representation" allows us to uniquely compose any symmetric state of n qubits $|\psi\rangle_s$ from a sum over all permutations $P : S_N \rightarrow S_N$ of n undistinguishable qubits $\{|\phi_1\rangle, \ldots, |\phi_n\rangle\}$:

$$|\psi\rangle_s = K^{-1/2} \sum_{\text{perm}} |\phi_{P(1)}\rangle |\phi_{P(2)}\rangle \cdots |\phi_{P(n)}\rangle \quad , \text{ with} \tag{11}$$

$$|\phi_i\rangle = \cos\frac{\theta_i}{2}|0\rangle + e^{i\varphi_i} \sin\frac{\theta_i}{2}|1\rangle \quad ,$$

and where the normalization factor K depends on the given state. The identity (11) allows the visualization of the multi-qubit state $|\psi\rangle_\mathrm{s}$ by n points on a sphere. In the following these points will be called the *Majorana points* (MP), and the sphere on which they lie the *Majorana sphere*.

As outlined in the previous section, for $n \geq 3$ qubits every closest product state $|\Lambda\rangle$ of a symmetric state $|\psi\rangle_\mathrm{s}$ is symmetric itself [26], and can therefore be written as $|\Lambda\rangle = |\sigma\rangle^{\otimes n}$, with a single qubit state $|\sigma\rangle$. The closest product states of a given symmetric state can therefore be visualized by Bloch vectors too, and in analogy to the Majorana points, we call $|\sigma\rangle$ a *closest product point* (CPP).

For symmetric states the scalar product from the definition of the geometric measure can be concisely expressed in terms of the MPs and a CPP:

$$|\langle \Lambda|\psi\rangle_\mathrm{s}| = n! \, K^{-1/2} \prod_{i=1}^{n} |\langle\sigma|\phi_i\rangle| \ . \tag{12}$$

To determine the CPP of a given symmetric state, one therefore has to maximize the absolute value of a product of scalar products. The factors $\langle\sigma|\phi_i\rangle$ are the angles between the corresponding Bloch vectors on the Majorana sphere, thus turning the determination of the CPP into a geometrical optimization problem.

From (11) it follows that the application on an arbitrary single-qubit unitary operation U to each of the n subsystems of a symmetric state $|\psi\rangle_\mathrm{s}$ yields

$$|\varphi\rangle_\mathrm{s} = U^{\otimes n}|\psi\rangle_\mathrm{s} = K^{-1/2} \sum_{\mathrm{perm}} \left(U|\phi_{P(1)}\rangle\right) \otimes \left(U|\phi_{P(2)}\rangle\right) \otimes \cdots \otimes \left(U|\phi_{P(n)}\rangle\right) \ . \tag{13}$$

Thus the symmetric state $|\psi\rangle_\mathrm{s}$ is mapped to a symmetric state $|\varphi\rangle_\mathrm{s}$ whose MP distribution can be obtained from a joint rotation of the MPs of $|\psi\rangle_\mathrm{s}$ along a common axis on the Majorana sphere. The two LOCC-equivalent states $|\psi\rangle_\mathrm{s}$ and $|\varphi\rangle_\mathrm{s}$ have the same *relative* MP distribution, and therefore the same number and relative distribution of CPPs, as well as the same amount of entanglement.

4.1 Examples

For pure symmetric states of two qubits the only absolute degree of freedom in the Majorana representation (and hence entanglement) is the distance between the two MPs. It is easy to determine that the CPP lies halfway between the two MPs, and that the entanglement is maximized when the MPs lie antipodal to each other. Figure 1 (a) shows the Bell state $|\psi^+\rangle = 1/\sqrt{2}\,(|01\rangle + |10\rangle)$ with its two MPs $|\phi_1\rangle = |0\rangle$ and $|\phi_2\rangle = |1\rangle$. Due to this azimuthal symmetry the CPPs form a continuous ring $|\sigma\rangle = 1/\sqrt{2}\,(|0\rangle + \mathrm{e}^{\mathrm{i}\varphi}|1\rangle)$, with $\varphi \in [0, 2\pi)$ around the equator. The amount of entanglement is $E_\mathrm{g}(|\psi^+\rangle) = 1$.

For three qubits the GHZ state and W state, two positive symmetric states, are considered to be extremal [56], with the W state proven to be the maximally entangled state in terms of the geometric measure [10].

The MPs of the tripartite GHZ state $|\mathrm{GHZ}\rangle = 1/\sqrt{2}\,(|000\rangle + |111\rangle)$ are, up to normalization,

$$|\phi_1\rangle = |0\rangle + |1\rangle \ , \quad |\phi_2\rangle = |0\rangle + \mathrm{e}^{\mathrm{i}2\pi/3}|1\rangle \ , \quad |\phi_2\rangle = |0\rangle + \mathrm{e}^{\mathrm{i}4\pi/3}|1\rangle \ . \tag{14}$$

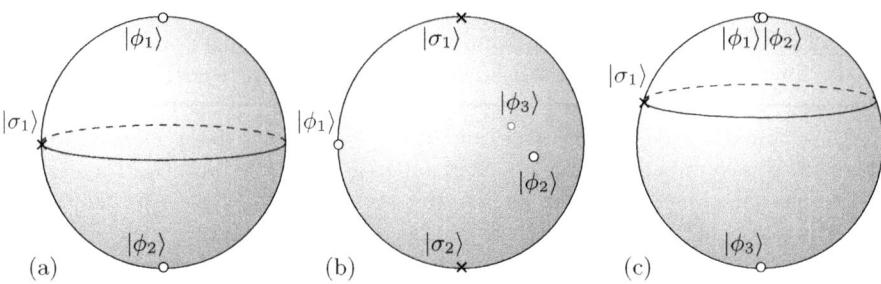

Fig. 1. Majorana representations of symmetric states of two and three qubits. MPs are depicted as white dots and CPPs as crosses or dashed lines. The pictures show (a) the two qubit Bell state $|\psi^+\rangle$, (b) three qubit GHZ state and (c) three qubit W state.

Its two CPPs are $|\sigma_1\rangle = |0\rangle$ and $|\sigma_2\rangle = |1\rangle$, and the amount of entanglement is $E_g(|\text{GHZ}\rangle) = 1$. Figure 1(b) shows the Majorana representation of the GHZ state. The three MPs form an equilateral triangle on the equator, and the two CPPs are the north pole and south pole.

In the case of the W state $|\text{W}\rangle = |S_{3,1}\rangle = 1/\sqrt{3}\,(|001\rangle + |010\rangle + |100\rangle)$, a Dicke state, the MPs can be directly accessed from its definition as $|\phi_1\rangle = |\phi_2\rangle = |0\rangle$ and $|\phi_3\rangle = |1\rangle$. The positive CPP follows from (5) as $|\sigma_1\rangle = \sqrt{2/3}\,|0\rangle + \sqrt{1/3}\,|1\rangle$, and the azimuthal symmetry implies that the set of all CPPs is formed by $|\sigma\rangle = \sqrt{2/3}\,|0\rangle + e^{i\varphi}\sqrt{1/3}\,|1\rangle$, with $\varphi \in [0, 2\pi)$. The Majorana representation is shown in Fig. 1(c), and the entanglement is $E_g(|\text{W}\rangle) = \log_2(9/4) \approx 1.17$.

4.2   Extremal Point Distributions

With (12) the min-max-problem (2) of finding the maximally entangled symmetric state can be recast as

$$\min_{\{|\phi_i\rangle\}} K^{-1/2}\left(\max_{|\sigma\rangle} \prod_{i=1}^{n} |\langle\sigma|\phi_i\rangle|\right). \tag{15}$$

Solving this "Majorana problem" is far from trivial, particularly with the normalization factor K depending on the MPs. The problem can be understood as an optimization problem on the sphere, prompting the question whether the known solutions of classical point distribution problems on the sphere [66] can help in finding the solutions of the Majorana problem. Two problems that have been extensively studied in the past are Tóth's problem and Thomson's problem.

Tóth's problem states that n points have to be distributed over the sphere so that the minimum pairwise distance becomes maximal [66]. Point configurations that solve this problem are known as spherical codes or sphere packings.

Thomson's problem is considering n point charges which are confined to the surface of a sphere and interacting with each other through Coulomb's inverse square law. The desired distribution is the one which minimizes the potential energy [57]. This problem has a variety of applications, e.g. for multi-electron bubbles in liquid Helium [35], liquid metal drops confined in Paul traps [12], shell structure of viruses [42], colloidosomes [15], fullerene patterns [32] and Abrikosov lattice of vortices in superconducting metal shells [16].

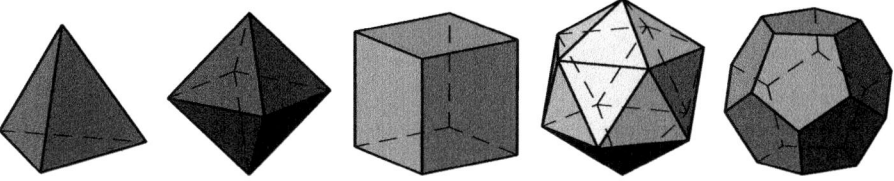

Fig. 2. (color online) The five Platonic solids from left to right: tetrahedron ($n = 4$), octahedron ($n = 6$), cube ($n = 8$), icosahedron ($n = 12$), and dodecahedron ($n = 20$)

Exact solutions to Tóth's and Thomson's problem of n points are known only for very few and low n [18,66], but numerical solutions are known for a much wider range of n in both problems [1,2]. An illustrating example are the five Platonic solids – the regular convex polyhedra whose edges, vertices and angles are all congruent, see Fig. 2. Because of their high symmetry one would expect that their vertices solve Tóth's and Thomson's problem for the corresponding n. This is however true only for $n = 4, 6, 12$, but not for $n = 8, 20$.

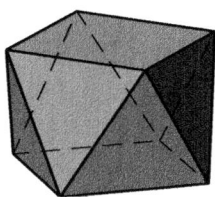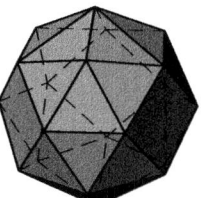

Fig. 3. (color online) For $n = 8$ the solution of Tóth's problem is given by a cubic antiprism, and for $n = 20$ by a polyhedron consisting of 30 triangles and 3 rhombuses

Figure 3 depicts the polyhedra that solve Tóth's problem for $n = 8, 20$. For $n = 8$ the solution is the cubic antiprism, which can be obtained from the cube by rotating one face by 45 degrees, followed by a slight compression along the direction perpendicular to the rotated face. In this way, the nearest neighbor distances between the vertices can be equally reduced, at the expense of breaking the high Platonic symmetry. This simple example shows that it is in general not easy to find the optimal spherical distribution for a set of points, and this is also true for the Majorana problem.

5   Analytic Results about MPs and CPPs

This section summarizes the analytic results about the Majorana representation that we have presented in [4]. In particular, the relationship between the coefficients of a symmetric state $|\psi\rangle_s = \sum_{k=0}^{n} a_k |S_k\rangle$ and the corresponding distribution of MPs and CPPs on the Majorana sphere will be illuminated.

Theorem 2. *For every symmetric n qubit state $|\psi\rangle_s$ the following holds:*

$$\int_0^{2\pi} \int_0^{\pi} |\langle \lambda(\theta, \varphi)|\psi\rangle_s|^2 \sin\theta \, \mathrm{d}\theta \mathrm{d}\varphi = \frac{4\pi}{n+1} \ , \tag{16}$$

where $|\lambda(\theta, \varphi)\rangle = \left(\cos\frac{\theta}{2}|0\rangle + \mathrm{e}^{\mathrm{i}\varphi} \sin\frac{\theta}{2}|1\rangle\right)^{\otimes n}$.

For the proof of this theorem we refer to [4]. The remarkable property of (16) is that the integral is the same for all symmetric n qubit states, thus straightforwardly yielding the upper bound $E_g^{\max} \leq \log_2(n+1)$ on the maximal symmetric entanglement. The integrand of (16) can be visualized by a spherical plot, and the constant integration volume can be understood as the constant volume of the plot. Figure 7(b) shows such a plot for a symmetric 12 qubit state.

Majorana representations with a high degree of symmetry are particularly easy to investigate. It is therefore elucidating to know the necessary and sufficient conditions for a rotational symmetry of the MP distribution.

Lemma 3. *The MP distribution of a symmetric n qubit state $|\psi\rangle_s$ is rotationally symmetric around the Z-axis with rotational angle $\theta = \frac{2\pi}{m}$ ($1 < m \leq n$) iff*

$$\forall\{k_i, k_j | a_{k_i} \neq 0 \wedge a_{k_j} \neq 0\} : (k_i - k_j) \bmod m = 0 \ . \tag{17}$$

This lemma states that all non-vanishing coefficients must be spaced apart from each other by a multiple of $m > 1$. An example of a rotationally symmetric state with $\theta = \pi/2$ would be $|\psi\rangle_s = a_3|S_3\rangle + a_7|S_7\rangle + a_{15}|S_{15}\rangle$.

Symmetric states whose coefficients are all real can be associated with a reflective symmetry of the Majorana representation along the X-Z-plane. From a mathematical point of view two Bloch vectors $|\phi_1\rangle$ and $|\phi_2\rangle$ exhibit such a reflective symmetry iff they are complex conjugates, i.e. $|\phi_1\rangle = \cos\frac{\theta}{2}|0\rangle + \mathrm{e}^{\mathrm{i}\varphi} \sin\frac{\theta}{2}|1\rangle$ and $|\phi_2\rangle = \cos\frac{\theta}{2}|0\rangle + \mathrm{e}^{-\mathrm{i}\varphi} \sin\frac{\theta}{2}|1\rangle = |\phi_1\rangle^*$.

Lemma 4. *Let $|\psi\rangle_s$ be a symmetric state of n qubits. $|\psi\rangle_s$ is real iff all its MPs are reflective symmetric with respect to the X-Z-plane of the Majorana sphere.*

It immediately follows from (15) that this reflective symmetry is also inherited to the CPPs.

Particularly strong results about the number and locations of CPPs can be obtained for positive symmetric states. With the exception of the Dicke states, any positive symmetric state can have at most $2n - 4$ CPPs, and it is believed that this result also holds for general symmetric states. Dicke states are a special case due to their continuous azimuthal symmetry, resulting in an uncountable number of CPPs.

Lemma 5. *Let $|\psi\rangle_s$ be a positive symmetric state of n qubits, excluding the Dicke states.*

(a) *If $|\psi\rangle_s$ is not rotationally symmetric around the Z-axis, then all its CPPs are positive.*

(b) *If $|\psi\rangle_s$ is rotationally symmetric around the Z-axis with minimal rotational angle $\frac{2\pi}{m}$, then all its CPPs $|\sigma(\theta,\varphi)\rangle = \cos\frac{\theta}{2}|0\rangle + e^{i\varphi}\sin\frac{\theta}{2}|1\rangle$ are restricted to the m azimuthal angles given by $\varphi = \varphi_r = \frac{2\pi r}{m}$ with $r \in \mathbb{Z}$. Furthermore, if $|\sigma(\theta,\varphi_r)\rangle$ is a CPP for some r, then it is also a CPP for all other values of r.*

The restriction of the CPPs to certain azimuthal angles imposed by this lemma is crucial for the rather technical proof (c.f. Appendix B of [4]) of the following statement about the number and locations of the CPPs.

Theorem 6. *The Majorana representation of every positive symmetric state of n qubits, excluding the Dicke states, belongs to one of the following three classes.*

(a) *$|\psi\rangle_s$ is rotationally symmetric around the Z-axis, with only the two poles as possible CPPs.*

(b) *$|\psi\rangle_s$ is rotationally symmetric around the Z-axis, with at least one CPP being non-positive.*

(c) *$|\psi\rangle_s$ is not rotationally symmetric around the Z-axis, and all CPPs are positive.*

Regarding the CPPs of states from class (b) and (c), the following assertions can be made for $n \geq 3$ qubits:

(b) *If both poles are occupied by at least one MP each, then there are at most $2n - 4$ CPPs, else there are at most n CPPs.*

(c) *There are at most $\lceil \frac{n+2}{2} \rceil$ CPPs*

The upper bound on the number of CPPs is intriguing, because the Euler characteristic implies that convex polyhedra with n vertices have at most $2n - 4$ faces. One could therefore ask whether there exists a deeper relationship between the CPPs and the faces of the MP distribution.

6   Solutions for Up to Twelve Qubits

An exhaustive search for the maximally entangled symmetric state over the whole space of symmetric states becomes infeasible already for only a few qubits, because the min-max problem (15) is too intractable to easily determine solutions. The results from the previous section as well as the fact that the maximally entangled state must have at least two CPPs (c.f. Lemma 4 in [4]) considerably simplify the numerical search for high and maximal symmetric entanglement, particularly among the subset of positive symmetric states, allowing the reliable determination of the maximally entangled positive symmetric states of up to

12 qubits. For the general non-positive case an exhaustive search over the entire Hilbert space is still too involved, so we concentrated on sets of promising states. Such states include those with highly spread out MP distributions and those that share qualitative features with the solutions to the classical optimization problems. Table 1 summarizes the presumed values of maximal geometric entanglement for symmetric states in the positive and general case. For comparison purposes, the known upper and lower bounds are also listed. For a detailed presentation and discussion of all the solutions we refer to [4].

Table 1. Values for the maximal entanglement of symmetric n qubit states in terms of the geometric measure. The entanglement values listed are (from left to right) those of the most entangled Dicke state, the maximally entangled positive symmetric state, the presumably maximally entangled symmetric state and the upper bound on symmetric entanglement. The relation $E_g(|S_{\lfloor n/2 \rfloor}\rangle) \leq E_g(|\Psi_n^{\text{pos}}\rangle) \leq E_g(|\Psi_n\rangle) < \log_2(n+1)$ holds for all n, and wherever the amount of entanglement does not increase, the respective right-hand table cell has been intentionally left blank. All numerical values have been calculated for ten or more digits, and the dagger $\dagger$ in the second column indicates values whose analytic form is known, but not displayed due to their complicated form.

n	$E_g(|S_{\lfloor n/2 \rfloor}\rangle)$	$E_g(|\Psi_n^{\text{pos}}\rangle)$	$E_g(|\Psi_n\rangle)$	$\log_2(n+1)$
2	1			$\log_2 3$
3	$\log_2(9/4)$			2
4	$\log_2(8/3)$	$\log_2 3$		$\log_2 5$
5	$\approx 1.532\,824\,877$	$\approx 1.742\,268\,948\ ^\dagger$		$\approx 2.584\,962\,501$
6	$\log_2(16/5)$	$\log_2(9/2)$		$\log_2 7$
7	$\approx 1.767\,313\,935$	$\approx 2.298\,691\,396\ ^\dagger$		3
8	$\approx 1.870\,716\,983$	$\approx 2.445\,210\,159$		$\approx 3.169\,925\,001$
9	$\approx 1.942\,404\,615$	$\approx 2.553\,960\,277\ ^\dagger$		$\approx 3.321\,928\,095$
10	$\approx 2.022\,720\,077$	$\approx 2.679\,763\,092$	$\approx 2.737\,432\,003$	$\approx 3.459\,431\,619$
11	$\approx 2.082\,583\,285$	$\approx 2.773\,622\,669$	$\approx 2.817\,698\,505$	$\approx 3.584\,962\,501$
12	$\approx 2.148\,250\,959$	$\approx 2.993\,524\,700$	$\log_2(243/28)$	$\approx 3.700\,439\,718$

For $n = 2, 3$ qubits the maximally entangled states were already identified as the Bell states and the W state, respectively. For $n = 4, 6, 12$ the Majorana problem is solved by the respective Platonic solid, i.e. the MP distributions are given by the vertices of the corresponding Platonic solid.

The "tetrahedron state" of four qubits, shown in Fig. 4, has the form $|\Psi_4\rangle = 1/\sqrt{3}\,|S_0\rangle + \sqrt{2/3}\,|S_3\rangle$. Since the state is positive and has a Z-axis rotational symmetry, Lemma 5 restricts the CPPs to the three half-circles shown as blue lines in Fig. 4(a). By means of the tetrahedral rotation group it is possible to find a unitary operation $U \neq \mathbb{1}$ so that (13) maps $|\Psi_4\rangle$ onto itself. This can be understood as a rotation on the Majorana sphere which moves each MP to the location of another MP. A rotation of this type, with the Bloch vector of $|\phi_4\rangle$ acting as the rotation axis, is performed twice between Fig. 4(a) and Fig. 4(c). For each of these configurations Lemma 5 gives rise to separate restrictions on

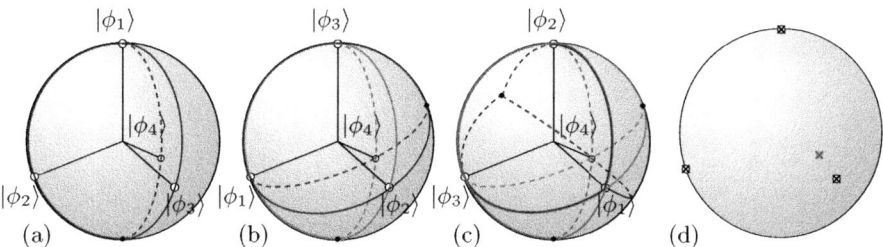

Fig. 4. The CPPs of the positive symmetric tetrahedron state $|\Psi_4\rangle$ of four qubits can be directly obtained from the tetrahedral rotation group and Lemma 5. Applying finite rotations amounts to permutations of the MPs and thus additional restrictions for the locations of the CPPs are obtained from Lemma 5.

the locations of the CPPs, and the intersection of all these restrictions leaves only four points, the MPs themselves. Therefore $|\Psi_4\rangle$ has four CPPs which coincide with the MPs.

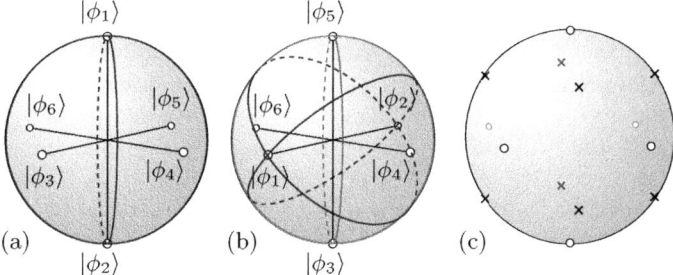

Fig. 5. Only one rotation from the octahedral rotation group is required to uniquely determine the locations of the eight CPPs of the octahedron state $|\Psi_6\rangle$

For the "octahedron state" of six qubits $|\Psi_6\rangle = 1/\sqrt{2}(|S_1\rangle + |S_5\rangle)$, shown in Fig. 5, the CPPs can be determined in the same way. Only one rotation from the octahedral rotation group is required to find the eight CPPs at the intersections of the blue and green lines depicted in Fig. 5(b). The CPPs lie at the center of each face of the octahedron, forming a cube inside the Majorana sphere. In contrast to the tetrahedron state with its overlapping MPs and CPPs, the CPPs of the octahedron state lie as far away from the MPs as possible. This is because (15) would be zero if a CPP $|\sigma\rangle$ were to lie antipodal to a MP $|\phi_i\rangle$.

For five points the solution to the classical problems is the trigonal bipyramid [2], and the corresponding "trigonal bipyramid state" $|\psi_5\rangle = 1/\sqrt{2}(|S_1\rangle + |S_4\rangle)$ is shown in Fig. 6(a). This is however not the maximally entangled symmetric state, and a numerical search yields the "square pyramid state" $|\Psi_5\rangle \approx 0.547|S_0\rangle + 0.837|S_4\rangle$, shown in Fig. 6(b), as the maximally entangled one. All its MPs and CPPs can be determined analytically by solving quartic equations. One of the five CPPs coincides with the north pole while the other four are equidistantly spread over a horizontal plane in the southern hemisphere. Notably, the "center

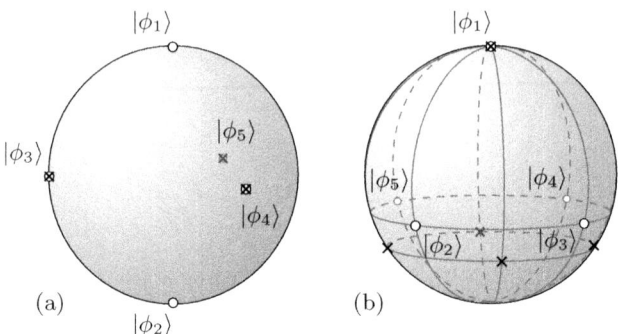

Fig. 6. The "trigonal bipyramid state" is shown in (a), but the Majorana problem of five qubits is solved by the "square pyramid state" shown in (b)

of mass" of the five MPs of $|\Psi_5\rangle$ does not coincide with the origin of the sphere, and the implications of this will be outlined in Sect. 6.2.

There is strong evidence that the "icosahedron state" $|\Psi_{12}\rangle = \sqrt{7}\,|S_1\rangle - \sqrt{11}\,|S_6\rangle - \sqrt{7}\,|S_{11}\rangle$, shown in Fig. 7(a), is the maximally entangled symmetric state of 12 qubits. The MPs form the vertices of a regular icosahedron, while the 20 CPPs are centered on the faces of the icosahedron, describing a dodecahedron inside the Majorana sphere. Figure 7(b) is the spherical plot of the function $f(\theta, \varphi) = |\langle\lambda(\theta, \varphi)|\Psi_{12}\rangle|$ which already appeared as the integrand of (16). This function is variously known as the characteristic polynomial, Majorana polynomial [29], amplitude function [51] or coherent state decomposition [34]. The CPPs and MPs of a symmetric state can be readily identified as the global maxima and the antipodes of the zeros of $f(\theta, \varphi)$, respectively.

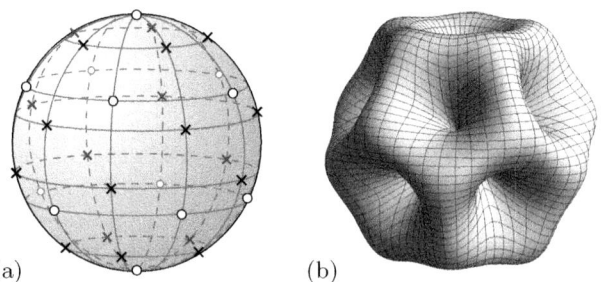

Fig. 7. The MPs and CPPs of the 12 qubit "icosahedron state" $|\Psi_{12}\rangle$ are depicted in (a), and the corresponding amplitude function $f(\theta, \varphi) = |\langle\lambda(\theta, \varphi)|\Psi_{12}\rangle|$ is shown in (b). For $|\Psi_{12}\rangle$ the locations of the MPs and CPPs coincide with the zeros and maxima of $f(\theta, \varphi)$, respectively.

6.1 Dual Polyhedra

Each of the five Platonic solids shown in Fig. 2 has a dual polyhedron with faces and vertices interchanged, and this dual polyhedron is again a Platonic solid [64]. As seen in Fig. 8, the octahedron and cube form a dual pair, and so do the icosahedron and dodecahedron. In contrast to this, the tetrahedron is self-dual, i.e. it is its own dual.

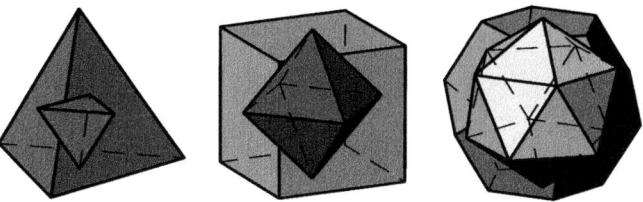

Fig. 8. (color online) The relationships between the Platonic solids and their duals

Interestingly, these dualities are also inherited to the Majorana representations of the corresponding symmetric quantum states. For example, we have seen that the 20 CPPs of the icosahedron state $|\Psi_{12}\rangle$ form the vertices of a dodecahedron. On the other hand, when considering the 20 qubit "dodecahedron state" $|\Psi_{20}\rangle = \sqrt{187}|S_0\rangle + \sqrt{627}|S_5\rangle + \sqrt{247}|S_{10}\rangle - \sqrt{627}|S_{15}\rangle + \sqrt{187}|S_{20}\rangle$, it is easy to show that this state has 12 CPPs which occupy the vertices of an icosahedron. Thus the Majorana representation of the dodecahedron state can be immediately obtained from Fig. 7 (a) by interchanging the MPs and CPPs. The same duality exists between the octahedron state and the cube state, c.f. Fig. 5 (c). Furthermore, the tetrahedron state is its own dual, as seen in Fig. 4 (d). Unlike the dual of the Platonic solid, however, the dual tetrahedron state is not turned "upside down" as seen in Fig. 8, but rather coincides with the original tetrahedron state.

6.2 Anticoherent Spin States and the Queens of Quantum

As outlined in Sect. 4, there exists an isomorphism between the states of a spin-j particle and the symmetric states of $2j$ qubits. The coherent states of a quantum particle can be regarded as the most classical states, and in terms of the Majorana representation these states are those whose MPs all coincide at a single point, thus describing a "classical" spin vector. Anticoherent spin states, first studied in [69], are states that exhibit maximally nonclassical behavior in the sense that their spin vector vanishes. Since such states can be considered the "opposite" of coherent states, it would be interesting to determine the MPs and the geometric entanglement of their symmetric counterparts. For example, one could ask whether maximally entangled symmetric states correspond to anticoherent states or to the mathematical concept of spherical designs [11]. However, the fact that the "center of mass" of the five qubit square pyramid state $|\Psi_5\rangle$ does not coincide with the origin of the Majorana sphere straightforwardly implies that this state is neither anticoherent nor a spherical design[2].

An alternative to anticoherent states was formulated in [19], where the least classical states are coined "queens of quantum". The Majorana representations of these states differ from our maximally entangled symmetric states, but when replacing the Hilbert-Schmidt metric with the Bures metric [41], the solutions of the two problems become identical. In other words, the Majorana representation of the spin-j "queen of quantum" in terms of the Bures metric is identical to that of the maximally entangled symmetric state of $2j$ qubits in terms of the geometric measure.

7   Conclusion

We have analyzed and discussed the geometric entanglement of highly and maximally entangled symmetric states of n qubits. The upper bound on symmetric entanglement rules out symmetric states as exact, deterministic MBQC resources. For the case of approximate MBQC we present arguments against the usefulness of symmetric states, and provide a proof for the class of Dicke states. With the known analytic results about the Majorana representation of symmetric states it is easy to numerically determine the most entangled states and to discuss their properties. As an example we showed how the determination of the CPPs of "Platonic states" is greatly simplified with the help of the theoretical results. With the help of the maximally entangled symmetric five qubit state it was shown that the solutions to the Majorana problem do not necessarily relate to anticoherent states or spherical designs. It is found that the well-known concept of the dual polyhedra of Platonic solids possesses a direct analog for symmetric quantum states, thereby deepening the relationship between the Majorana representation and the polyhedra of classical geometry.

Acknowledgments. The authors would like to thank S. Miyashita, A. Soeda, S. Virmani, K.-H. Borgwardt and M. Van den Nest for very helpful discussions. This work is supported by the National Research Foundation & Ministry of Education, Singapore and the project "Quantum Computation: Theory and Feasibility" in the framework of the CNRS-JST Strategic French-Japanese Cooperative Program on ICT. MM thanks the "Special Coordination Funds for Promoting Science and Technology" for financial support.

References

1. Altschuler, E.L., Williams, T.J., Ratner, E.R., Dowla, F., Wooten, F.: Phys. Rev. Lett. 72, 2671 (1994)
2. Ashby, N., Brittin, W.E.: Am. J. Phys. 54, 776 (1986)

---

[2] For anticoherent spin states this readily follows from $\langle \Psi_5 | S_z | \Psi_5 \rangle \neq 0$. For spherical designs we observe that by setting $p(x) = x$ in Definition 2 of [11], it follows that for all spherical designs the "center of mass" must necessarily coincide with the sphere's origin.

3. Aulbach, M., Markham, D. (in preparation)
4. Aulbach, M., Markham, D., Murao, M.: New J. Phys. 12, 073025 (2010)
5. Bacry, H.: J. Math. Phys. 15, 1686 (1974)
6. Barnett, R., Turner, A., Demler, E.: Phys. Rev. A 76, 013605 (2007)
7. Barnum, H., Linden, N.: J. Phys. A: Math. Gen. 34, 6787 (2001)
8. Bastin, T., Krins, S., Mathonet, P., Godefroid, M., Lamata, L., Solano, E.: Phys. Rev. Lett. 103, 070503 (2009)
9. Cavalcanti, D.: Phys. Rev. A 73, 044302 (2006)
10. Chen, L., Xu, A., Zhu, H.: Phys. Rev. A 82, 032301 (2010)
11. Crann, J., Pereira, R., Kribs, D.W.: J. Phys. A: Math. Theo. 43, 255307 (2010)
12. Davis, E.J.: Aerosol Sci. Technol. 26, 212 (1997)
13. D'Hondt, E., Panangaden, P.: Quant. Inf. Comp. 6, 173 (2006)
14. Dicke, R.H.: Phys. Rev. 93, 99 (1954)
15. Dinsmore, A.D., Hsu, M.F., Nikolaides, M.G., Marquez, M., Bausch, A.R., Weitz, D.A.: Science 298, 1006 (2002)
16. Dodgson, M.J.W., Moore, M.A.: Phys. Rev. B 55, 3816 (1997)
17. Dür, W., Vidal, G., Cirac, J.I.: Phys. Rev. A 62, 062314 (2000)
18. Erber, T., Hockney, G.M.: J. Phys. A: Math. Gen. 24, L1369 (1991)
19. Giraud, O., Braun, P.A., Braun, D.: New J. Phys. 12, 063005 (2010)
20. Gross, D., Flammia, S.T., Eisert, J.: Phys. Rev. Lett. 102, 190501 (2009)
21. Hannay, J.H.: J. Phys. A: Math. Gen. 101, L101 (1996)
22. Hannay, J.H.: J. Phys. A: Math. Gen. 31, L53 (1998)
23. Hayashi, M., Markham, D., Murao, M., Owari, M., Virmani, S.: Phys. Rev. Lett. 96, 040501 (2006)
24. Hayashi, M., Markham, D., Murao, M., Owari, M., Virmani, S.: Phys. Rev. A 77, 012104 (2008)
25. Horodecki, R., Horodecki, P., Horodecki, M., Horodecki, K.: Rev. Mod. Phys. 81, 865 (2009)
26. Hübener, R., Kleinmann, M., Wei, T.C., González-Guillén, C., Gühne, O.: Phys. Rev. A 80, 032324 (2009)
27. Jung, E., Hwang, M.R., Kim, H., Kim, M.S., Park, D., Son, J.W., Tamaryan, S.: Phys. Rev. A 77, 62317 (2008)
28. Kolenderski, P.: Open Systems & Information Dynamics 17, 107 (2009)
29. Kolenderski, P., Demkowicz-Dobrzanski, R.: Phys. Rev. A 78, 052333 (2008)
30. Korbicz, J.K., Cirac, J.I., Lewenstein, M.: Phys. Rev. Lett. 95, 120502 (2005)
31. Korbicz, J.K., Gühne, O., Lewenstein, M., Häffner, H., Roos, C.F., Blatt, R.: Phys. Rev. A 74, 052319 (2006)
32. Kroto, H.W., Heath, J.R., O'Brien, S.C., Curl, R.F., Smalley, R.E.: Nature 318, 162 (1985)
33. De Lathauwer, L., De Moor, B., Vandewalle, J.: SIAM J. Matrix Anal. Appl. 21, 1324 (2000)
34. Leboeuf, P.: J. Phys. A: Math. Gen. 24, 4575 (1991)
35. Leiderer, P.: Z. Phys. B 98, 303 (1995)
36. Majorana, E.: Nuovo Cimento 9, 43–50 (1932)
37. Mäkelä, H., Suominen, K.A.: Phys. Rev. Lett. 99, 190408 (2007)
38. Markham, D. (2010), arXiv:1001.0343
39. Markham, D., Miyake, A., Virmani, S.: New J. Phys. 9, 194 (2007)
40. Markham, D., Vedral, V.: Phys. Rev. A 67, 042113 (2003)
41. Martin, J., Giraud, O., Braun, P.A., Braun, D., Bastin, T.: Phys. Rev. A 81, 062347 (2010)

42. Marzec, C.J., Day, L.A.: Biophys. J. 65, 2559 (1993)
43. Mathonet, P., Krins, S., Godefroid, M., Lamata, L., Solano, E., Bastin, T.: Phys. Rev. A 81, 052315 (2010)
44. Mora, C.E., Piani, M., Miyake, A., Van den Nest, M., Dür, W., Briegel, H.J.: Phys. Rev. A 81, 042315 (2010)
45. Van den Nest, M., Dür, W., Miyake, A., Briegel, H.J.: New J. Phys. 9, 204 (2007)
46. Van den Nest, M., Miyake, A., Dür, W., Briegel, H.J.: Phys. Rev. Lett. 97, 150504 (2006)
47. Ni, G., Wang, Y.: Math. Comput. Modelling 46, 1345 (2007)
48. Orús, R., Dusuel, S., Vidal, J.: Phys. Rev. Lett. 101, 025701 (2008)
49. Plenio, M.B., Virmani, S.: Quant. Inf. Comp. 7, 1 (2007)
50. Prevedel, R., Cronenberg, G., Tame, M.S., Paternostro, M., Walther, P., Kim, M.S., Zeilinger, A.: Phys. Rev. Lett. 103, 020503 (2009)
51. Radcliffe, J.M.: J. Phys. A: Math. Gen. 4, 313 (1971)
52. Renner, R.: Ph.D. thesis, ETH Zurich (2005), arXiv:quant-ph/0512258
53. Ribeiro, P., Vidal, J., Mosseri, R.: Phys. Rev. E 78, 021106 (2008)
54. Shimony, A.: Ann. NY. Acad. Sci. 755, 675 (1995)
55. De Silva, V., Lim, L.H.: SIAM J. Matrix Anal. Appl. 30, 1084 (2008)
56. Tamaryan, S., Wei. T.C., Park, D.: Phys. Rev. A 80, 052315 (2009)
57. Thomson, J.J.: Phil. Mag. 7, 237 (1904)
58. Tóth, G.: J. Opt. Soc. Am. B 24, 275 (2007)
59. Tóth, G., Gühne, O.: Phys. Rev. Lett. 102, 170503 (2009)
60. Vedral, V., Plenio, M.B.: Phys. Rev. A 57, 1619 (1998)
61. Vidal, G., Tarrach, R.: Phys. Rev. A 59, 141 (1999)
62. Wei, T.C., Ericsson, M., Goldbart, P.M., Munro, W.J.: Quant. Inf. Comp. 4, 252 (2004)
63. Wei, T.C., Goldbart, P.M.: Phys. Rev. A 68, 042307 (2003)
64. Wenninger, M.J.: Dual Models. Cambridge University Press, Cambridge (1983)
65. Werner, R.F., Holevo, A.S.: J. Math. Phys. 43, 4353 (2002)
66. Whyte, L.L.: Am. Math. Mon. 59, 606 (1952)
67. Wieczorek, W., Krischek, R., Kiesel, N., Michelberger, P., Tóth, G., Weinfurter, H.: Phys. Rev. Lett. 103, 020504 (2009)
68. Zhu, H., Chen, L., Hayashi, M.: New J. Phys. 12, 083002 (2010)
69. Zimba, J.: Electron. J. Theor. Phys. 3, 143 (2006)
70. Zimba, J., Penrose, R.: Stud. Hist. Phil. Sci. 24, 697–720 (1993)

Monogamy of Multi-qubit Entanglement in Terms of Rényi and Tsallis Entropies

Jeong San Kim and Barry C. Sanders

Institute for Quantum Information Science, University of Calgary, Alberta T2N 1N4,
Canada
jekim@ucalgary.ca, sandersb@ucalgary.ca
http://www.iqis.org

Abstract. We summarize our recent result about monogamy of multi-qubit entanglement: Using Rényi-α entropy, we provide a class of monogamy inequalities of multi-qubit entanglement for $\alpha \geq 2$. We also provide another class of monogamy inequalities in terms of Tsallis-q entropy for $2 \leq q \leq 3$.

Keywords: Rényi entropy, Tsallis entropy, monogamy of entanglement.

1 Introduction

One distinct property of quantum entanglement from other classical correlations is its restricted sharability. For instance, if a pair of parties in a multipartite quantum system share maximal entanglement, then they can share neither entanglement [1,2] nor classical correlations [3] with the rest. This is known as the *Monogamy of Entanglement* (MoE) [4], and it has been shown that this restricted sharability of quantum entanglement can be used as a resource to distribute a secret key which is secure against unauthorized parties [5,6].

Whereas MoE is a restricted property of entanglement in multipartite quantum systems, the sharability itself is about the bipartite entanglements among the parties in multipartite systems. In other words, it is inevitable to have a proper way of quantifying bipartite entanglement for a good description of the monogamy nature in multipartite quantum systems. For this reason, certain criteria of bipartite entanglement measure were recently proposed for a good description of the monogamy nature of entanglement in multipartite quantum systems [7]; that is,

(i) *Monotonicity*: the property that ensures entanglement cannot be increased under local operations and classical communications.
(ii) *Separability*: capability of distinguishing entanglement from separability.
(iii) *Monogamy*: upper bound on a sum of bipartite entanglement measures thereby showing that bipartite sharing of entanglement is bounded.

The first mathematical characterization of MoE was shown in three-qubit systems [1] using *concurrence* [8] as the bipartite entanglement measure. It is

W. van Dam et al. (Eds.): TQC 2010, LNCS 6519, pp. 159–167, 2011.

known as *CKW-inequality* named after its establishers, Coffman, Kundu and Wootters, and this CKW-type inequality was also shown for arbitrary multi-qubit systems later [2]. In other words, concurrence is a good entanglement measure for multi-qubit systems that satisfies the criteria proposed in [7].

However, monogamy inequality using concurrence is know to fail in its generalization for higher-dimensional quantum systems [9,7]. Furthermore, although MoE in multi-qubit systems is mathematically well- characterized in terms of concurrence, it is not generally true for other entanglement measures such as *Entanglement of Formation* (EoF) [10]. In other words, MoE does not have CKW-type characterization in terms of EoF, and this exposes the importance of the choice of a bipartite entanglement measure to characterize MoE even in multi-qubit systems. Moreover, for possible generalization of monogamy inequality into higher-dimensional quantum systems, it is undoubtedly one of the most important and necessary tasks to have a proper way of quantifying bipartite entanglement.

Rényi-α entropy [11] and Tsallis-q entropy are two most representative generalizations of Shannon entropy [12] in terms of the non-negative real parameter α and q, and they have been widely used in the study of quantum information theory such as quantum entanglement and correlations [13,14,15]. For the case when α and q tend to 1, Rényi-α and Tsallis-q entropies converge to Shannon entropy.

Here, we consider the full spectrums of Rényi-α and Tsallis-q entropies with respect to the parameters α and q, and show that they can also provide monogamy inequalities of multi-qubit entanglement for a various choice of the parameters α and q. Although EoF that is based on Shannon entropy is known to fail for usual CKW-type characterization of MoE, we show that Rényi-α and Tsallis-q entropies can still have CKW-type monogamy inequality for all case of $\alpha \geq 2$ and $2 \leq q \leq 3$.

This paper is organized as follows. In Section 2.1, we recall the definition of quantum Rényi-α and Tsallis-q entropies, and define bipartite entanglement measures namely *Rényi-α entanglement* and *Tsallis-q entanglement*. In Section 2.2, we provide an analytic formula of Rényi-α entanglement and Tsallis-q entanglement for arbitrary two-qubit states for $\alpha \geq 1$ and $1 \leq q \leq 4$. In Section 3, we derive the monogamy inequalities of multi-qubit entanglement in terms of Rényi-α entanglement for $\alpha \geq 2$ and Tsallis-q entanglement for $2 \leq q \leq 3$. Finally, we summarize our results in Section 4.

2 Rényi-α Entropy and Entanglement Measures

2.1 Definition

For any quantum state ρ, its quantum Rényi-α entropy is defined as

$$S_\alpha(\rho) = \frac{1}{1-\alpha} \log \mathrm{tr}\rho^\alpha, \tag{1}$$

for any $\alpha > 0$ and $\alpha \neq 1$ [16], and its Tsallis-q entropy is defined as

$$T_q(\rho) = \frac{1}{q-1}\left(1 - \mathrm{tr}\rho^q\right), \tag{2}$$

for any $q > 0$ and $q \neq 1$. In the limiting case that $\alpha, q \rightarrow 1$, it can be easily checked that $S_\alpha(\rho)$ and $T_q(\rho)$ converge to the von Neumann entropy, that is

$$\lim_{\alpha \rightarrow 1} S_\alpha(\rho) = S(\rho) = \lim_{q \rightarrow 1} T_q(\rho). \tag{3}$$

In other words, Rényi-α and Tsallis-q entropies have a singularity at $\alpha, q = 1$, and it is removable by von Neumann entropy. Here, we will just consider $S_1(\rho) = T_1(\rho) = S(\rho)$ for any quantum state ρ.

For a bipartite pure state $|\psi\rangle_{AB}$, the von Neumann entropy of the reduced density matrix $\rho_A = \mathrm{tr}_B|\psi\rangle_{AB}\langle\psi|$ is known to be a good bipartite entanglement measure

$$E(|\psi\rangle_{AB}) = S(\rho_A) = S(\rho_B). \tag{4}$$

With noticing that von Neumann entropy quantifies the uncertainty of the quantum state, this way of quantifying bipartite entanglement is based on the uncertainty of subsystem: More uncertainty on subsystems implies stronger quantum correlation between subsystems.

A well-known way to generalize this concept of entanglement measure into mixed states is taking the minimum (or infimum) of the average entanglements

$$E_\mathrm{f}(\rho_{AB}) = \min \sum_i p_i E(|\psi\rangle_{AB}) \tag{5}$$

over all possible pure state decompositions of the mixed state $\rho_{AB} = \sum_i p_i |\psi_i\rangle_{AB}\langle\psi_i|$. This generalization is known as *convex-roof extension*, and $E_\mathrm{f}(\rho_{AB})$ is called the entanglement of formation of ρ_{AB}.

As a generalization of EoF into the full spectrum of Rényi-α entropy [17], *Rényi-α entanglement* of a bipartite pure state $|\psi\rangle_{AB}$ is defined as

$$\mathcal{R}_\alpha(|\psi\rangle_{AB}) = S_\alpha(\rho_A), \tag{6}$$

where $\rho_A = \mathrm{tr}_B|\psi\rangle_{AB}\langle\psi|$, and for a mixed state ρ_{AB}, its Rényi-α entanglement is defined as,

$$\mathcal{R}_\alpha(\rho_{AB}) = \min \sum_i p_i \mathcal{R}_\alpha(|\psi_i\rangle_{AB}), \tag{7}$$

where the minimum is taken over all possible pure state decompositions of $\rho_{AB} = \sum_i p_i |\psi_i\rangle_{AB}\langle\psi_i|$.

Let us also define *Tsallis-q entanglement* as

$$\mathcal{T}_q(|\psi\rangle_{AB}) := T_q(\rho_A), \tag{8}$$

for a pure state $|\psi\rangle_{AB}$, and

$$\mathcal{T}_q(\rho_{AB}) := \min \sum_i p_i \mathcal{T}_q(|\psi_i\rangle_{AB}), \tag{9}$$

for a mixed state ρ_{AB}.

Similar to EoF for bipartite quantum states, Rényi-α and Tsallis-q entanglements are based on the uncertainty of subsystems, which has EoF as a special case when $\alpha, q \to 1$.

It is direct to check that $\mathcal{R}_\alpha(\rho_{AB}) = 0$ if and only if ρ_{AB} is a separable state, and this also holds for $\mathcal{T}_q(\rho_{AB})$. Furthermore, we can show that Rényi-α and Tsallis-q entanglements are entanglement monotone: They are not increased under local quantum operations and classical communications. Thus both are good entanglement measures in bipartite quantum systems satisfying the first two criteria in [7].

2.2   Analytical Formula for Two-qubit Systems

Let us recall the definition of concurrence. For any bipartite pure state $|\phi\rangle_{AB}$, its concurrence, $\mathcal{C}(|\phi\rangle_{AB})$ is defined as [8]

$$C(|\phi\rangle_{AB}) = \sqrt{2(1 - \mathrm{tr}\rho_A^2)}, \tag{10}$$

where $\rho_A = \mathrm{tr}_B(|\phi\rangle_{AB}\langle\phi|)$, and for any mixed state ρ_{AB}, its concurrence is defined as

$$C(\rho_{AB}) = \min \sum_k p_k C(|\phi_k\rangle_{AB}), \tag{11}$$

where the minimum is taken over all possible pure state decompositions, $\rho_{AB} = \sum_k p_k |\phi_k\rangle_{AB}\langle\phi_k|$.

For any two-qubit mixed state ρ_{AB}, its concurrence is known to have an analytic formula [8], that is,

$$C(\rho_{AB}) = \max\{0, \lambda_1 - \lambda_2 - \lambda_3 - \lambda_4\}, \tag{12}$$

where λ_i's are the eigenvalues, in decreasing order, of $\sqrt{\sqrt{\rho_{AB}}\tilde{\rho}_{AB}\sqrt{\rho_{AB}}}$ and $\tilde{\rho}_{AB} = \sigma_y \otimes \sigma_y \rho_{AB}^* \sigma_y \otimes \sigma_y$ with the Pauli operator σ_y. Furthermore, the relation between concurrence and EoF of a two-qubit mixed state ρ_{AB} (or a pure state $|\psi\rangle_{AB}$ in $2 \otimes d$ systems), can be given as a monotone increasing, convex function $\mathcal{E}$ [8], such that

$$E_{\mathrm{f}}(\rho_{AB}) = \mathcal{E}(\mathcal{C}_{AB}), \tag{13}$$

where

$$\mathcal{E}(x) = H\left(\frac{1}{2} + \frac{1}{2}\sqrt{1 - x^2}\right), \quad \text{for } 0 \leq x \leq 1, \tag{14}$$

with the binary entropy function $H(t) = -t \log t - (1 - t) \log(1 - t)$. In other words, the analytic formula of concurrence as well as its functional relation with EoF lead us to an analytic formula of EoF for two-qubit states.

For any two-qubit pure state (or any pure state with Schmidt-rank less than or equal to two) with its Schmidt decomposition $|\psi\rangle_{AB} = \sqrt{\lambda_0}|00\rangle_{AB} + \sqrt{\lambda_1}|11\rangle_{AB}$, it can be easily checked that

$$\mathcal{R}_\alpha(|\psi\rangle_{AB}) = \frac{1}{1 - \alpha} \log(\lambda_0^\alpha + \lambda_1^\alpha), \tag{15}$$

and

$$\mathcal{T}_q\left(|\psi\rangle_{AB}\right) = \frac{1}{q-1}\left(1 - \lambda_0^q - \lambda_1^q\right). \tag{16}$$

Now, for each $\alpha, q > 0$, by defining analytic functions

$$f_\alpha(x) := \frac{1}{1-\alpha}\log\left[\left(\frac{1-\sqrt{1-x^2}}{2}\right)^\alpha + \left(\frac{1+\sqrt{1-x^2}}{2}\right)^\alpha\right] \tag{17}$$

and

$$g_q(x) := \frac{1}{q-1}\left[1 - \left(\frac{1+\sqrt{1-x^2}}{2}\right)^q - \left(\frac{1-\sqrt{1-x^2}}{2}\right)^q\right] \tag{18}$$

on $0 \leq x \leq 1$, we have

$$\mathcal{R}_\alpha\left(|\psi\rangle_{AB}\right) = f_\alpha\left(\mathcal{C}(|\psi\rangle_{AB})\right), \tag{19}$$

and

$$\mathcal{T}_q\left(|\psi\rangle_{AB}\right) = g_q\left(\mathcal{C}(|\psi\rangle_{AB})\right). \tag{20}$$

where $\mathcal{C}(|\psi\rangle_{AB})$ is the concurrence of $|\psi\rangle_{AB}$. Thus, for each $\alpha, q > 0$, we have functional relations between Rényi-α and Tsallis-q entanglements and concurrence for pure states with Schmidt-rank 2. Here, we note that $f_\alpha(x)$ and $g_q(x)$ converge to the function $\mathcal{E}(x)$ in Eq. (14) for the case when α and q tend to 1.

For two-qubit mixed states, it was shown that there exists an optimal decomposition for the concurrence of a two-qubit mixed state such that every pure state concurrence in the decomposition has the same value [8]. Based on this, one possible sufficient condition for the relations in Eqs. (19) and (20) to be also true for mixed states is that the functions $f_\alpha(x)$ and $g_q(x)$ are monotonically increasing and convex [18]. In other words, we have

$$\mathcal{R}_\alpha\left(\rho_{AB}\right) = f_\alpha\left(\mathcal{C}(\rho_{AB})\right), \quad \mathcal{T}_q\left(\rho_{AB}\right) = g_q\left(\mathcal{C}(\rho_{AB})\right) \tag{21}$$

for any two-qubit mixed state ρ_{AB} provided that $f_\alpha(x)$ and $g_q(x)$ are monotonically increasing and convex. Moreover, for the ranges of α and q where $f_\alpha(x)$ and $g_q(x)$ are monotonically increasing and convex, Eq. (21) also implies analytic formulas of Rényi-α and Tsallis-q entanglements entanglement respectively, for any two-qubit state.

Theorem 1. *For any real $\alpha \geq 1$,*

$$f_\alpha(x) = \frac{1}{1-\alpha}\log\left[\left(\frac{1-\sqrt{1-x^2}}{2}\right)^\alpha + \left(\frac{1+\sqrt{1-x^2}}{2}\right)^\alpha\right] \tag{22}$$

is a monotonically increasing and convex function for $0 \leq x \leq 1$, and so is

$$g_q(x) = \frac{1}{q-1}\left[1 - \left(\frac{1+\sqrt{1-x^2}}{2}\right)^q - \left(\frac{1-\sqrt{1-x^2}}{2}\right)^q\right] \tag{23}$$

for $1 \leq q \leq 4$.

Although the proof of Theorem 1 is analytically, it contains a few complicated calculations. The complete proof can be found in [18,19].

3   Entanglement Constraint in Multi-party Quantum Systems

Using concurrence as the bipartite entanglement quantification, the monogamous property of a multi-qubit pure state $|\psi\rangle_{A_1 A_2 \cdots A_n}$ was shown to have a mathematical characterization as,

$$\mathcal{C}^2_{A_1(A_2 \cdots A_n)} \geq \mathcal{C}^2_{A_1 A_2} + \cdots + \mathcal{C}^2_{A_1 A_n}, \tag{24}$$

where $\mathcal{C}_{A_1(A_2 \cdots A_n)} = \mathcal{C}(|\psi\rangle_{A_1 A_2 \cdots A_n})$ is the concurrence of $|\psi\rangle_{A_1 A_2 \cdots A_n}$ with respect to the bipartite cut between A_1 and the others, and $\mathcal{C}_{A_1 A_i} = \mathcal{C}(\rho_{A_1 A_i})$ is the concurrence of the reduced density matrix $\rho_{A_1 A_i}$ for $i = 2, \ldots, n$ [1,2].

Here, we establish a mathematical formulation for the monogamous property of multi-qubit entanglement in terms of Rényi-α and Tsallis-q entanglement. Before this, we first note an important property of the functions $f_\alpha(x)$ and $g_q(x)$ Eqs. (19) and (20) that plays a crucial role in the proof of Rényi and Tsallis entanglement monogamy.

Lemma 1. *For any real $\alpha \geq 2$, we have*

$$f_\alpha\left(\sqrt{x^2 + y^2}\right) \geq f_\alpha(x) + f_\alpha(y) \tag{25}$$

for $0 \leq x, \ y \leq 1$ such that $0 \leq x^2 + y^2 \leq 1$. Similarly, we have

$$g_q\left(\sqrt{x^2 + y^2}\right) \geq g_q(x) + g_q(y), \tag{26}$$

for $2 \leq q \leq 3$.

The proof of Lemma 1 is analytical, however it contains a few complicated calculations. The complete proof can be found in [18,19].

Now, by using Theorem 1 together with Lemma 1, we have the following theorem, which is the monogamy inequality of multi-qubit entanglement in terms of Rényi and Tsallis entanglement.

Theorem 2. *For $\alpha \geq 2$, $2 \leq q \leq 3$ and any multi-qubit mixed state $\rho_{A_1 A_2 \cdots A_n}$, we have*

$$\mathcal{R}_\alpha\left(\rho_{A_1(A_2 \cdots A_n)}\right) \geq \mathcal{R}_\alpha\left(\rho_{A_1 A_2}\right) + \cdots + \mathcal{R}_\alpha\left(\rho_{A_1 A_n}\right), \tag{27}$$

and

$$\mathcal{T}_q\left(\rho_{A_1(A_2 \cdots A_n)}\right) \geq \mathcal{T}_q\left(\rho_{A_1 A_2}\right) + \cdots + \mathcal{T}_q\left(\rho_{A_1 A_n}\right), \tag{28}$$

where $\mathcal{R}_\alpha\left(\rho_{A_1(A_2 \cdots A_n)}\right)$ and $\mathcal{T}_q\left(\rho_{A_1(A_2 \cdots A_n)}\right)$ are the Rényi-α and Tsallis-q entanglements of $\rho_{A_1 A_2 \cdots A_n}$ with respect to the bipartite cut between A_1 and the others respectively, and $\mathcal{R}_\alpha\left(\rho_{A_1 A_i}\right)$ and $\mathcal{R}_\alpha\left(\rho_{A_1 A_i}\right)$ are the Rényi-α and and Tsallis-q entanglements of the reduced density matrix $\rho_{A_1 A_i}$ on two-qubit subsystem $A_1 A_i$ respectively for $i = 2, \ldots, n$.

Proof. Here, we provide the prove of Rényi-α monogamy inequality of entanglement, and the proof for Tsallis-q entanglement is then analogous.

Let us first consider an n-qubit pure state $|\psi\rangle_{A_1 A_2 \cdots A_n}$. From Eq. (24),

$$\mathcal{C}_{A_1(A_2\cdots A_n)} \geq \sqrt{\mathcal{C}^2_{A_1 A_2} + \cdots + \mathcal{C}^2_{A_1 A_n}}, \tag{29}$$

where $\mathcal{C}_{A_1(A_2\cdots A_n)}$ and $\mathcal{C}_{A_1 A_i}$ are the concurrences of $|\psi\rangle_{A_1(A_2\cdots A_n)}$ and $\rho_{A_1 A_i}$ for each $i = 2, \cdots n$, respectively. Because $|\psi\rangle_{A_1(A_2\cdots A_n)}$ has Schmidt-rank less than or equal to two (it is a $2 \otimes 2^{\otimes n-1}$ pure state with respect to the bipartition between A$_1$ and the rest), its concurrence and Rényi-α entanglement are related by the function $f_\alpha(x)$ in Eq. (19); that is,

$$\mathcal{R}_\alpha\left(|\psi\rangle_{A_1(A_2\cdots A_n)}\right) = f_\alpha\left(\mathcal{C}_{A_1(A_2\cdots A_n)}\right). \tag{30}$$

Now we have

$$\begin{aligned}
\mathcal{R}_\alpha\left(|\psi\rangle_{A_1(A_2\cdots A_n)}\right) &= f_\alpha\left(\mathcal{C}_{A_1(A_2\cdots A_n)}\right) \\
&\geq f_\alpha\left(\sqrt{\mathcal{C}^2_{A_1 A_2} + \cdots + \mathcal{C}^2_{A_1 A_n}}\right) \\
&\geq f_\alpha\left(\mathcal{C}_{A_1 A_2}\right) + f_\alpha\left(\sqrt{\mathcal{C}^2_{A_1 A_3} + \cdots + \mathcal{C}^2_{A_1 A_n}}\right) \\
&\vdots \\
&\geq f_\alpha\left(\mathcal{C}_{A_1 A_2}\right) + \cdots + f_\alpha\left(\mathcal{C}_{A_1 A_n}\right) \\
&= \mathcal{R}_\alpha\left(\rho_{A_1 A_2}\right) + \cdots + \mathcal{R}_\alpha\left(\rho_{A_1 A_n}\right),
\end{aligned} \tag{31}$$

where the first inequality follows from the monotonicity of $f_\alpha(x)$, the second inequality is due to Eq. (25) by letting $x = \mathcal{C}_{A_1 A_2}$ and $y = \sqrt{\mathcal{C}^2_{A_1 A_3} + \cdots + \mathcal{C}^2_{A_1 A_n}}$, the other inequalities are from iterative use of Eq. (25), and the last equality follows from the functional relation between Rényi-α entanglement and concurrence for two-qubit states.

For an n-qubit mixed state $\rho_{A_1 A_2 \cdots A_n}$, let $\rho_{A_1(A_2\cdots A_n)} = \sum_j p_j |\psi_j\rangle_{A_1(A_2\cdots A_n)}\langle\psi_j|$ be an optimal decomposition for $\mathcal{R}_\alpha\left(\rho_{A_1(A_2\cdots A_n)}\right)$ such that $\mathcal{R}_\alpha\left(\rho_{A_1(A_2\cdots A_n)}\right) = \sum_j p_j \mathcal{R}_\alpha\left(|\psi_j\rangle_{A_1(A_2\cdots A_n)}\right)$. Because each $|\psi_j\rangle_{A_1(A_2\cdots A_n)}$ in the decomposition is an n-qubit pure state, the monogamy inequality in terms of Rényi-α entanglement holds for each $|\psi_j\rangle_{A_1(A_2\cdots A_n)}$. Thus, we have

$$\begin{aligned}
\mathcal{R}_\alpha\left(\rho_{A_1(A_2\cdots A_n)}\right) &= \sum_j p_j \mathcal{R}_\alpha\left(|\psi_j\rangle_{A_1(A_2\cdots A_n)}\right) \\
&\geq \sum_j p_j\left(\mathcal{R}_\alpha\left(\rho^j_{A_1 A_2}\right) + \cdots + \mathcal{R}_\alpha\left(\rho^j_{A_1 A_n}\right)\right) \\
&= \sum_j p_j \mathcal{R}_\alpha\left(\rho^j_{A_1 A_2}\right) + \cdots + \sum_j p_j \mathcal{R}_\alpha\left(\rho^j_{A_1 A_n}\right) \\
&\geq \mathcal{R}_\alpha\left(\rho_{A_1 A_2}\right) + \cdots + \mathcal{R}_\alpha\left(\rho_{A_1 A_n}\right),
\end{aligned} \tag{32}$$

where $\rho_{A_1 A_i}^j$ is the reduced density matrix of $|\psi_j\rangle_{A_1(A_2\cdots A_n)}$ onto subsystem $A_1 A_i$ for each $i = 2, \cdots, n$ and the last inequality is by definition of Rényi-α entanglement for each $\rho_{A_1 A_i}$.

4   Conclusion

Using Rényi-α entropy and Tsallis-q entropy, we have established a class of monogamy inequalities of multi-qubit entanglement. We have shown that monogamy of multi-qubit entanglement can have CKW-type characterization in terms of Rényi-α entanglement for $\alpha \geq 2$, and Tsallis-q entropy for $2 \leq q \leq 3$.

Multipartite entanglement is known to have many inequivalent classes, which are not convertible to each other under *Stochastic Local operations and classical communications* (SLOCC) [20]. Furthermore, the number of inequivalent classes increases dramatically as the number of parties increase [21]. Not like bipartite entanglement, the existence of inequivalent classes of multipartite entanglement implies that the states from different classes are hardly comparable to each other in such a way of comparing a single parameter that quantifies their entanglement. This is one of the main difficulties in the study of multipartite entanglement.

Whereas the interconvertibility of quantum states under SLOCC gives us an operational way to classify multipartite entanglement, entangled states from different classes can also reveal different characters with respect to their monogamy and polygamy properties. For example, three-qubit systems are known to have two inequivalent classes of genuine three-qubit entanglement, the Greenberger-Horne-Zeilinger (GHZ) class [22] and the W-class [20]. In terms of monogamy and polygamy relations, CKW and its dual inequalities are saturated by W-class states, while the differences between terms in the inequalities can assume their largest values for GHZ-class states. In other words, monogamy and polygamy of multipartite entanglement can also be used for an analytical characterization of entanglement in multipartite quantum systems.

The class of monogamy inequalities of multi-qubit entanglement we provided here consists of infinitely many inequalities parameterized by α and q. We believe that this selective choice of our monogamy inequalities will leads us to an efficient way of analytic classification of multi-qubit entanglement. Moreover, our result will also provide useful tools and strong candidates for general monogamy relations of entanglement in multipartite higher-dimensional quantum systems, which is one of the most important and necessary topics in the study of multipartite quantum entanglement.

Acknowledgments

This work was supported by *i*CORE, MITACS and USARO. BCS is a CIFAR Fellow.

References

1. Coffman, V., Kundu, J., Wootters, W.K.: Phys. Rev. 61, 052306 (2000)
2. Osborne, V., Verstraete, F.: Phys. Rev. Lett. 96, 220503 (2006)
3. Koashi, M., Winter, A.: Phys. Rev. A 69, 022309 (2004)
4. Terhal, B.M.: IBM J. Research and Development 48, 71 (2004)
5. Renes, J.M., Grassl, M.: Phys. Rev. A 74, 022317 (2006)
6. Masanes, L.: Phys. Rev. Lett. 102, 140501 (2009)
7. Kim, J.S., Das, A., Sanders, B.C.: Phys. Rev. A 79, 012329 (2009)
8. Wootters, W.K.: Phys. Rev. Lett. 80, 2245 (1998)
9. Ou, Y.C.: Phys. Rev. A 75, 034305 (2007)
10. Bennett, C.H., DiVincenzo, D.P., Smolin, J.A., Wootters, W.K.: Phys. Rev. A 54, 3824 (1996)
11. Rényi, A.: Proceedings of the Fourth Berkeley Symposium on Mathematics, Statistics and Probability, p. 547. Berkeley University Press, Berkeley (1960)
12. Shannon, C.E.: The Bell System Technical Journal 30, 50–64 (1951)
13. Bovino, F.A., Castagnoli, G., Ekert, A., Horodecki, P., Alves, C.M., Sergienko, A.V.: Phys. Rev. Lett. 95, 240407 (2005)
14. Terhal, B.M.: J. Theor. Comp. Sci. 287, 313 (2002)
15. Lévay, P., Nagy, S., Pipek, J.: Phys. Rev. A 72, 022302 (2005)
16. Horodecki, R., Horodecki, P., Horodecki, M.: Phys. Lett. A 210, 377 (1996)
17. Vidal, G.: J. Mod. Opt. 47, 355 (2000)
18. Kim, J.S., Sanders, B.C.: Journal of Physics A: Mathematical and Theoretical 43(44), 442305 (2010)
19. Kim, J.S.: Phys. Rev. A 81, 062328 (2010)
20. Dür, W., Vidal, G., Cirac, J.I.: Phys. Rev. A 62, 062314 (2000)
21. Osterloh, A., Siewert, J.: Phys. Rev. A 72, 012337 (2005)
22. Greenberger, D.M., Horne, M.A., Zeilinger, A.: Bell's Theorem, Quantum Theory, and Conceptions of the Universe. In: Kafatos, M. (ed.), p. 69. Kluwer, Dordrecht (1989)

Bypassing State Initialisation in Perfect State Transfer Protocols on Spin-Chains

Carlo Di Franco[1], Mauro Paternostro[2], and M.S. Kim[3]

[1] Department of Physics, University College Cork, Cork, Republic of Ireland
[2] School of Mathematics and Physics, Queen's University, Belfast BT7 1NN,
United Kingdom
[3] Institute for Mathematical Sciences, Imperial College London, SW7 2PG,
United Kingdom
and
QOLS, The Blackett Laboratory, Imperial College London, Prince Consort Road,
SW7 2BW, United Kingdom

Abstract. Although a complete picture of the full evolution of complex quantum systems would certainly be the most desirable goal, for particular Quantum Information Processing schemes such an analysis is not necessary. When quantum correlations between only specific elements of a many-body system are required for the performance of a protocol, a more distinguished and specialised investigation is helpful. Here, we provide a striking example with the achievement of perfect state transfer in a spin chain without state initialisation, whose realisation has been shown to be possible in virtue of the correlations set between the first and last spin of the transmission-chain.

1 Introduction

Quantum Information Theory (QIT) is having a remarkable impact from a fundamental viewpoint by providing alternative perspectives to physical problems using new conceptual instruments. The study of quantum correlations shared by many distinctive objects is helping us in understanding their behaviour at critical points [1] and quantifying the resources required in order to efficiently simulate such situations [2]. The simulation of complex quantum systems is usually a prohibitive task for even the most powerful classical machine due to the exponential growth of its Hilbert space with respect to the number of elements. In this context, several advances have recently been made in the study of the ground state of particular many-particle systems [3]. While all the proposed methods have found use in simulating static properties of ground states, their application to the investigation of time evolution is, in general, problematic. However, although the analysis of the complete behaviour of quantum many-particle systems will be a fundamental task in QIT, for the study of some particular Quantum Information Processing (QIP) schemes this is not necessary. When quantum correlations between only specific elements of a many-body system are required for the performance of a protocol, a more distinguished and specialised approach is helpful. The problems related to simulating the

W. van Dam et al. (Eds.): TQC 2010, LNCS 6519, pp. 168–174, 2011.

dynamics of many-particle systems can be solved, in this context, by considering not their whole evolution, but only the behaviour of a few characteristic features. Here we provide a striking example in the achievement of perfect Quantum State Transfer (QST) in a spin chain without state initialisation, whose realisation has been shown to be possible in virtue of the correlations set between the first and last spin of the transmission-chain [4].

This result is also important on a more pragmatic ground. Recently, it has been shown that the control over multipartite registers for the purposes of QIP can be sensibly reduced in a way so as to avoid the generally demanding fast and accurate inter-qubit switching and gating. In this case, the price to pay for the performance of efficient operations is the pre-engineering of appropriate patterns of couplings [5]. The preparation of a fiducial state for the initialisation of a QIP device can be however experimentally demanding. This is mainly due to the difficulty of preparing pure states of multipartite systems, which is one of DiVincenzo's criteria [6], a set of requirements that any QIP system should meet. Remarkably, our proposal is able to bypass the initialisation of the spin-medium in a known pure state. The scheme requires only end-chain single-qubit operations and a single application of a global unitary evolution and is thus fully within a scenario where the control over the core part of the spin medium is relaxed in favour of controllability of the first and last element of the chain. The relaxation of the conditions necessary for manipulating information is a fundamental step in order to shorten the time for the achievement of realistic QIP. This allows us to loose the requirements for information protection from environmental effects. Instead of utilising demanding always-on schemes for the shielding of the information content of a system, this could be done only during the running-time of the protocol.

2 Perfect State Transfer without State Initialisation for the XX Model

Spin chains have recently emerged as remarkable candidates for the realisation of faithful short-distance transmission of quantum information [7]. Here, the system under investigation is a nearest-neighbour XX coupling involving N spin-1/2 particles. Its Hamiltonian reads

$$\hat{\mathcal{H}} = \sum_{i=1}^{N-1} J_i(\hat{X}_i\hat{X}_{i+1} + \hat{Y}_i\hat{Y}_{i+1}), \tag{1}$$

where J_i is the interaction strength between spin i and $i+1$ and $\hat{X}$, $\hat{Y}$ and $\hat{Z}$ denote the x, y and z Pauli matrix, respectively. Let us start considering

$$J_i = J\sqrt{i(N-i)} \tag{2}$$

with J being a characteristic energy scale that depends on the specific physical implementation of the model (we choose units such that $\hbar = 1$ throughout the paper). This model has been extensively analysed [8]: $1 \to N$ perfect QST is achieved, through this coupling, when the initial state of all the spins but the first one is $|0\rangle$.

In our investigation, however, we drop the condition on the state of the central qubits, and we just assume control over the external ones. For the understanding of what follows, it is useful to analyse the time-evolution, in the Heisenberg picture, of the two-site operators $\hat{\mathbb{I}}_i \hat{Z}_{N-i+1}$, $\hat{X}_i \hat{X}_{N-i+1}$, and $\hat{X}_i \hat{Y}_{N-i+1}$. We define $\hat{O}(t)$ as the time-evolved form of a given operator $\hat{O}$. By solving a set of Heisenberg equations, we have that, at time $t^* = \pi/4J$ and for any N,

$$\hat{\mathbb{I}}_i(t^*)\hat{Z}_{N-i+1}(t^*) = \hat{Z}_i\,\hat{\mathbb{I}}_{N-i+1}. \tag{3}$$

On the other hand, for an even N we find

$$\hat{X}_i(t^*)\hat{X}_{N-i+1}(t^*) = \hat{X}_i \hat{X}_{N-i+1},$$
$$\hat{X}_i(t^*)\hat{Y}_{N-i+1}(t^*) = \hat{Y}_i \hat{X}_{N-i+1}, \tag{4}$$

while for an odd N

$$\hat{X}_i(t^*)\hat{X}_{N-i+1}(t^*) = \hat{Y}_i \hat{Y}_{N-i+1},$$
$$\hat{X}_i(t^*)\hat{Y}_{N-i+1}(t^*) = -\hat{X}_i \hat{Y}_{N-i+1}. \tag{5}$$

These results can also be easily obtained from the analysis presented in Ref. [9], where it is shown that the evolution of single-qubit operators can be evaluated by means of a method based on oriented graphs. For instance, in the case $N = 5$, $\hat{X}_1(t)$ can be decomposed as

$$\hat{X}_1(t) = \alpha_1(t)\hat{X}_1 + \alpha_2(t)\hat{Z}_1\hat{Y}_2 + \alpha_3(t)\hat{Z}_1\hat{Z}_2\hat{X}_3 +$$
$$\alpha_4(t)\hat{Z}_1\hat{Z}_2\hat{Z}_3\hat{Y}_4 + \alpha_5(t)\hat{Z}_1\hat{Z}_2\hat{Z}_3\hat{Z}_4\hat{X}_5. \tag{6}$$

Similarly,

$$\hat{X}_5(t) = \beta_1(t)\hat{X}_5 + \beta_2(t)\hat{Z}_5\hat{Y}_4 + \beta_3(t)\hat{Z}_5\hat{Z}_4\hat{X}_3 +$$
$$\beta_4(t)\hat{Z}_5\hat{Z}_4\hat{Z}_3\hat{Y}_2 + \beta_5(t)\hat{Z}_5\hat{Z}_4\hat{Z}_3\hat{Z}_2\hat{X}_1. \tag{7}$$

The time-evolution of the two-site operator $\hat{X}_1\hat{X}_5$ can be therefore obtained by considering the sum of all the possible products of elements of these two sets of operator. For $J_i = J\sqrt{i(N-i)}$, the time-dependent coefficients $\alpha_i(t)$ have the behaviour shown in Fig. 1. By symmetry, $\beta_i(t) = \alpha_i(t)$ for all values of i and t. It is easy to notice, in Fig. 1, that at the time $t^* = \pi/4J$, $\alpha_5(t^*) = \beta_5(t^*) = 1$, while all the other coefficients are equal to 0. For that particular time, therefore, the evolved operator $\hat{X}_1(t^*)\hat{X}_5(t^*)$ is just the product of $\hat{Z}_1\hat{Z}_2\hat{Z}_3\hat{Z}_4\hat{X}_5$ times $\hat{Z}_5\hat{Z}_4\hat{Z}_3\hat{Z}_2\hat{X}_1$. We have $\hat{X}_1(t^*)\hat{X}_5(t^*) = \hat{Y}_i\hat{Y}_{N-i+1}$. In the same way, all the other evolved operators in Eqs. (3) and (5) can be obtained.

Each of the two-site operators in Eqs. (3)-(5) evolves into operators acting on the same qubits, without any dependence on other operators of the chain. This paves the way to the core of our protocol, which we now describe qualitatively.

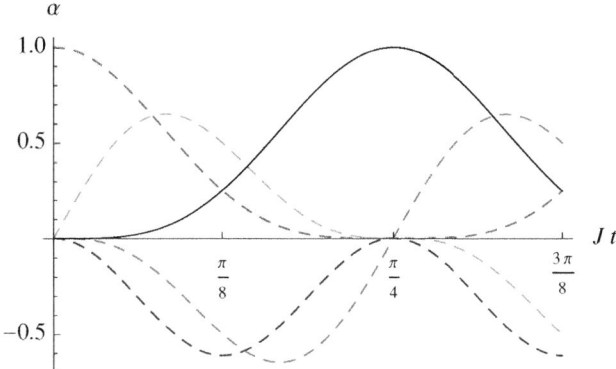

Fig. 1. Coefficients α_1 (red dashed line), α_2 (green dashed line), α_3 (blue dashed line), α_4 (purple dashed line) and α_5 (black line) against dimensionless time Jt, for $N = 5$ and $J_i = J\sqrt{i(N-i)}$.

Qubit 1 is initialised in the input state ρ^{in} (either a pure or mixed state) we want to transfer and qubit N is projected onto

$$|\pm_N\rangle = \frac{1}{\sqrt{2}}(|0\rangle \pm e^{iN\frac{\pi}{2}}|1\rangle). \tag{8}$$

In what follows, we say that outcome $+1$ (-1) is found if a projection onto $|+_N\rangle$ $(|-_N\rangle)$ is performed. Then the interaction encompassed by $\hat{\mathcal{H}}$ is switched on for a time $t^* = \pi/4J$, after which we end up with an entangled state of the chain. The amount of entanglement shared by the elements of the chain depends critically on their initial state. Regardless of the amount of entanglement being set, an $\hat{X}$-measurement over the first spin projects the N-th one onto a state that is locally-equivalent to ρ^{in}. More specifically, if the product of the measurement outcomes at 1 (after the evolution) and N (before the evolution)

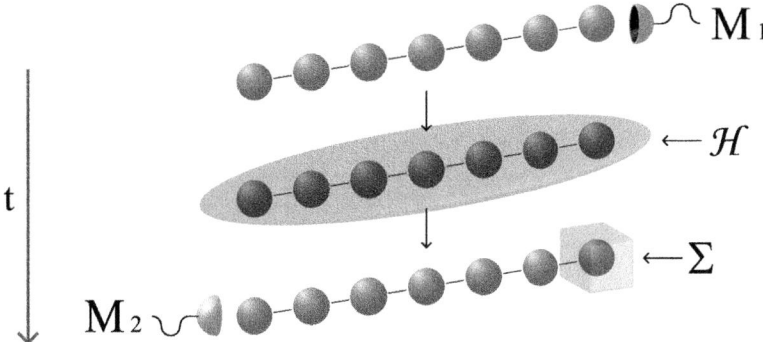

Fig. 2. Sketch of the scheme for perfect QST. M_1 and M_2 are measurements performed over a fixed basis and Σ is a conditional operation. $\hat{\mathcal{H}}$ is the Hamiltonian of the system.

is $+1$ (-1), the last spin will be in $(\hat{T}^N)^\dagger \rho^{in}(\hat{T}^N)$ $[\hat{Z}(\hat{T}^N)^\dagger \rho^{in}(\hat{T}^N)\hat{Z}]$, where $\hat{T} = |0\rangle\langle 0| + e^{i\frac{\pi}{2}}|1\rangle\langle 1|$ (therefore, $\hat{T}^2 = \hat{Z}$) [10]. In any case, apart from a simple single-spin transformation, perfect QST is achieved. A sketch of the scheme is presented in Fig. 2.

The crucial point here is that, regardless of the amount of entanglement established between the spin-medium and the extremal elements of the chain (*i.e.* spins 1 and N), upon $\hat{X}$-measurement of 1, the last spin is disconnected from the rest of the system, *whose initial state is inessential to the performance of the protocol* and could well be, for instance, a thermal state of the chain in equilibrium at finite temperature. In fact, the key requirements for our scheme are the arrangement of the proper time-evolution (to be accomplished within the coherence times of the system) and the performance of clean projective measurements on spin 1 and, preventively, on N.

3   General Conditions

In general, the protocol can be adapted to any Hamiltonian for which we can find a triplet of single-spin operators $\hat{B}, \hat{C}, \hat{D}$ such that, for symmetric spin pairs, we have

$$\hat{B}_i^{j_O}(t^*)\hat{C}_{N-i+1}\hat{O}_{N-i+1}(t^*) = \hat{O}_i\hat{D}_{N-i+1}^{k_O}. \tag{9}$$

Here, $\hat{B}_i$ ($\hat{D}_{N-i+1}$) provides the eigenbasis for the measurement over spin i ($N - i + 1$) of the chain after (before) the evolution, $\hat{C}_{N-i+1}$ is a decoding operation, $\hat{O}_i = \hat{O}_i(0) = \hat{X}, \hat{Y}, \hat{Z}$ and $j_O, k_O = 0, 1$, depending on the coupling model. We point out that, when these conditions are not fulfilled, our protocol can still be rather successful. In these cases, through an information flux approach, we can still estimate the average transfer fidelity [11]. For instance, we can consider the case in which we are able to engineer the strength of the coupling rates of just the extremal qubits (J_1 and J_{N-1}). Therefore, we take $J_i = J$ (for $i = 2, ..., N - 2$), $J_1 = J_{N-1} = \eta J$. The behaviour of this system against the dimensionless interaction time Jt and the inhomogeneity parameter η has already been studied in Ref. [11]. For simplicity, here we consider the time-dependent coefficients $\alpha_i(t)$ in the case $N = 5$, for the value of η which maximises QST fidelity ($\eta \sim 0.815$). Also this system is centro-symmetric, therefore we have $\beta_i(t) = \alpha_i(t)$ for all values of i and t. In this case, however, there is no time for which $\alpha_5(t) = \beta_5(t) = 1$, while all the other coefficients are equal to 0. Nevertheless, for a dimensionless time $Jt' \sim 1.9$, the value of α_5 and β_5 are close to 1, while all the other ones are close to 0. Our estimate gives an average transfer fidelity via our protocol of $F \sim \alpha_5^2(t') > 99.9\%$.

4   Remarks

We have shown the existence of a simple control-limited scheme for the achievement of perfect QST in a system of interacting spins without the necessity of demanding state initialisation. Our protocol requires just *one-shot* unitary evolution and end-chain local operations. Its efficiency arises from the establishment of *correlations* between the first and last spin of the transmission-chain. With the exception of limiting cases where the transfer is automatically achieved (as for the transfer of eigenstates of $\hat{X}_1$ when model $\hat{\mathcal{H}}$ is used), these are set regardless of the state of the spin medium, their amount being a case-dependent issue. The end-chain measurements, which are key to our scheme, "adjust" such correlations in a way so as to achieve perfect QST.

Due to the dependence of this protocol only on the correlations established between the first and last spin, the analysis of this scheme just requires the investigation of the time evolution of two-site operators. The exponential growth of the Hilbert space of the total state of the system does not affect our analysis, that can thus be done by means of only "slowly-growing" computational effort. In fact, the number of elements in the decomposition of the relevant two-site operators grows as N^2. Moreover, in this way, we were able to obtain our results removing any dependance on the state of all the central qubits.

We would like to conclude this contribution by remarking that our protocol for state transfer without initialisation is already encountering the attention of the community interested in quantum spin-chain dynamics. In fact, a recent proposal by Markiewicz and Wiesniak [12] has addressed a scheme for perfect state transfer without initialisation where the necessity for "remote coordination" between sending and receiving agents is bypassed.

Acknowledgments

We acknowledge support from the UK EPSRC. C.D.F. is supported by the Irish Research Council for Science, Engineering and Technology. M.P. thanks the UK EPSRC (EP/G004579/1) for financial support.

References

1. Osterloh, A., et al: Nature 416, 608 (2002); Osborne, T.J., Nielsen, M.A.: Phys. Rev. A 66, 032110 (2002)
2. Schuch, N., et al.: Phys. Rev. Lett. 100, 30504 (2008)
3. Wilson, K.G.: Rev. Mod. Phys. 47, 773 (1975); White, S.R.: Phys. Rev. Lett. 69, 2863 (1992); Fannes, M., Nachtergaele, B., Werner, R.F.: Lett. Math. Phys. 25, 249 (1992); Vidal, G.: Phys. Rev. Lett. 91, 147902 (2003); Verstraete, F., Garcia-Ripoll, J.J., Cirac, J.I.: Phys. Rev. Lett. 93, 207204 (2004); Schollwöck, U.: Rev. Mod. Phys. 77, 259 (2005); Verstraete, F., Cirac, J.I. (2004), e-print arXiv:cond-mat/0407066; Vidal, G.: Phys. Rev. Lett. 99, 220405 (2007); Anders, S., et al.: Phys. Rev. Lett. 97, 107206 (2006); Anders, S., Briegel, H.-J., Dür, W.: New J. Phys. 9, 361 (2007)
4. Di Franco, C., Paternostro, M., Kim, M.S.: Phys. Rev. Lett. 101, 230502 (2008)
5. Benjamin, S.C., Bose, S.: Phys. Rev. Lett. 90, 247901 (2003)
6. DiVincenzo, D.P.: Mesoscopic Electron Transport. In: Kowenhoven, L., Schön, G., Sohn, L. (eds.) Kluwer, Dordrecht (1997)

7. Bose, S.: Phys. Rev. Lett. 91, 207901 (2003); Bose, S.: Contemp. Phys. 48, 13 (2007); Burgarth, D.: PhD thesis, University College London (2006)

8. Christandl, M., et al.: Phys. Rev. Lett. 92, 187902 (2004); Christandl, M., et al.: Phys. Rev. A 71, 032312 (2005); Nikolopoulos, G.M., Petrosyan, D., Lambropoulos, P.: Europhys. Lett. 65, 297 (2004); J. Phys.: Condens. Matter 16, 4991 (2004)

9. Di Franco, C., Paternostro, M., Palma, G.M.: Int. J. Quant. Inf. 6(Supp. 1), 659 (2008)

10. The final state of spin N can be easily obtained from Eqs. (3)-(5). For instance, if N is even, $\langle \hat{Z}_N(t^*) \rangle = \langle \hat{Z}_1(0) \rangle$, $\langle \hat{X}_1 \hat{X}_N(t^*) \rangle = \langle \hat{X}_1 \hat{X}_N(0) \rangle$ and $\langle \hat{X}_1 \hat{Y}_N(t^*) \rangle = \langle \hat{Y}_1 \hat{X}_N(0) \rangle$. If qubit N has been projected onto $|\pm_N\rangle$ [for which $\langle \hat{X}_N(0) \rangle = \pm(-1)^{\frac{N}{2}}$], we have $\langle \hat{Z}_N(t^*) \rangle = \langle \hat{Z}_1(0) \rangle$, $\langle \hat{X}_1 \hat{X}_N(t^*) \rangle = \pm(-1)^{\frac{N}{2}} \langle \hat{X}_1(0) \rangle$ and $\langle \hat{X}_1 \hat{Y}_N(t^*) \rangle = \pm(-1)^{\frac{N}{2}} \langle \hat{Y}_1(0) \rangle$. The state of spin N, after the measurement performed on spin 1, will satisfy $\langle \hat{Z}_N(t^*) \rangle = \langle \hat{Z}_1(0) \rangle$, $\langle \hat{X}_N(t^*) \rangle = (-1)^{\frac{N}{2}} c \langle \hat{X}_1(0) \rangle$ and $\langle \hat{Y}_N(t^*) \rangle = (-1)^{\frac{N}{2}} c \langle \hat{Y}_1(0) \rangle$, where c is the product of the measurement outcomes at 1 (after the evolution) and N (before the evolution). The state $(\hat{Z}^{\frac{N}{2}}) \rho^{in} (\hat{Z}^{\frac{N}{2}})$ $[(\hat{Z}^{\frac{N}{2}+1}) \rho^{in} (\hat{Z}^{\frac{N}{2}+1})]$ satisfies these conditions for $c = 1$ ($c = -1$)

11. Di Franco, C., et al.: Phys. Rev. A 76, 042316 (2007)

12. Markiewicz, M., Wiesniak, M.: Phys. Rev. A 79, 054304 (2009)

Teleportation of a Quantum State of a Spatial Mode with a Single Massive Particle

Libby Heaney

Department of Physics, University of Oxford, Clarendon Laboratory,
Oxford, OX1 3PU, UK

Abstract. Mode entanglement exists naturally between regions of space in ultra-cold atomic gases. It has, however, been debated whether this type of entanglement is useful for quantum protocols. This is due to a particle number superselection rule that restricts the operations that can be performed on the modes. In this paper, we show how to exploit the mode entanglement of just a single particle for the teleportation of an unknown quantum state of a spatial mode. We detail how to overcome the superselection rule to create any initial quantum state and how to perform Bell state analysis on two of the modes. We show that two of the four Bell states can always be reliably distinguished, while the other two have to be grouped together due to an unsatisfied phase matching condition. The teleportation of an unknown state of a quantum mode thus only succeeds half of the time.

Keywords: Mode entanglement, quantum teleportation, superselection rule.

1 Introduction

Entanglement is a key resource in many practical applications using quantum mechanics [1]. Usually entanglement is thought to exist between the degrees of freedom of two, or more, well localised quanta, such as photons or massive particles. When particles are well separated, they are distinguishable from one another and can thus be assigned labels. Entanglement between the particles is then well defined since their Hilbert space has a tensor product structure. However, if the de Broglie wavelengths of identical particles begin to overlap, the particles become indistinguishable and one can no longer assign a label to each particle. The concept of standard particle entanglement breaks down as the Hilbert space no longer has the required tensor product structure; it is a projection onto the symmetric or antisymmetric subspaces (depending on the particle statistics) [2].

While entanglement between indistinguishable particles can still be correctly defined by taking a set of detectors into account [3], another method to recover a tensor product Hilbert space is to move into the formalism of second quantisation. Here, in the so called Fock basis, one defines a complete set of single particle states and counts the number of excitations in each. The corresponding mode

W. van Dam et al. (Eds.): TQC 2010, LNCS 6519, pp. 175–186, 2011.

structure has a tensor product Hilbert space and hence entanglement of modes is a meaningful concept. Entanglement can therefore exist between modes occupied by particles [2,4]. The modes can be energy eigenmodes or perhaps more relevant to quantum communication or information processing protocols are spatial modes. In the following section, we will discuss in more detail the mode structure for a simple many-body system. The particles may be massless, i.e. photons, or massive, such as those found in ultra-cold gases. Mode entanglement of photons has been considered in a number of works [5,6] and is usually limited to the single photon regime [7,8,9,10]. The experimental confirmation of the mode entanglement of a single photon via a Bell like test was obtained in 2004 [11] and the multipartite entanglement of one photon distributed between four optical modes was characterized using uncertainty relations in an interferometric setup in 2009 [12].

For massive particles, the existence of entanglement between spatial modes becomes less clear [13]. This is due to a particle number superselection rule [14,15], that forbids isolated systems from existing in a superposition of eigenstates of different mass. Hence, for modes occupied by massive particles the system density operator, $\hat{\rho}$, cannot contain any off-diagonal terms that connect states of different particle number. Any measurements made on the spatial modes are also restricted to the subspace of fixed particle number. Since the correlations of entanglement are locally basis independent, they can be confirmed via for instance a Bell inequality, where one should measure each subsystem locally in at least two basis. For spatial modes, the particle number is one such basis, however to measure in a second, rotated basis, measurements of superpositions of different numbers of particles are required. For an isolated system, such superposition measurements are forbidden due to the number of particles in the system being fixed. Until recently it was unresolved how to measure modes of a massive bosonic field in any way other than in the particle number basis. And this is why it has been debated [16] whether mode entanglement is as 'genuine' as particle entanglement or whether it is just a mathematical feature of the quantum state.

Recent research has, however, shown that superselection rules [17,18,19] can be overcome locally by using a suitable reference frame [17,20]. The reference frame required to rotate spatial modes away from the particle number basis is a coherent reservoir of particles such as a Bose-Einstein condensate (BEC). By using such a reservoir, one can create, at least in principle, superpositions of different numbers of particles [21,20]. Thus, it is predicted that mode entanglement of a single massive particle is, indeed, as genuine as particle entanglement [22]. Specific schemes for Bell inequality tests of mode entanglement of one [23,25,24], and more [26], massive particles have been given. Another question is whether mode entanglement can be used for quantum communication and quantum information processing. This is an interesting point, as mode entanglement occurs naturally in coherent ultra-cold bosonic gases [27,28,29,30,31]. If one could harness this entanglement for practical applications, it could by-pass the need to create complicated entangled states manually.

A first step to understanding the usefulness of the mode entanglement of just a single massive particle came in a recent paper [32], which gave a scheme for implementing the quantum dense coding protocol. Dense coding [33] allows the transmission of two classical bits of information via one qubit; in order to achieve the full quantum channel capacity a maximally entangled Bell state is initially required and full Bell state analysis is needed. It was shown [32] that the linear photonic dense coding channel capacity could be achieved without a BEC reservoir and that with a BEC reservoir the full quantum dense coding protocol could be implemented.

While it was briefly mentioned in [32] that the teleportation of a quantum state of a spatial mode should also be possible, no detailed scheme was given. It is the aim of this paper, to provide such a scheme. Quantum teleportation [34] is a key protocol in quantum information science and has been fruitfully demonstrated in experiments with a number of different physical implementations [35,36,37]. In [38], a qubit (or optical mode) consisting of the vacuum and one photon states was teleported using the entanglement of a single photon in a superposition of two spatial modes. Even though the creation of a superposition of different numbers of photons is not strictly forbidden by a superselection rule as it is with massive particles, there is still the problem of how to keep track of the phase between the two different photon number states. Lombardi *et al* [38] solved this by teleporting a mode that was entangled to another one. In other words, the single photon was coherently distributed across the mode to be teleported and an ancilla mode. This ancilla optical mode actually played the role of a phase reference in a similar way to how the BEC reservoir will play the role of a phase reference in our scheme here.

One motivation to study a teleportation scheme using the mode entanglement of a single massive particle is to allow for direct comparisons of quantum information processing with different physical systems. Moreover, from a fundamental viewpoint, it is interesting to clearly demonstrate that mode entanglement of massive particles is, in principle, useful entanglement - and can actually be used in a very similar way to particle entanglement - for quantum information processing despite the particle number superselection rule. In particular, in this paper we will see that in contrast to the dense coding scheme of [32], where the full channel capacity was attainable, here an arbitrary state of a quantum mode is only reliably teleported half of the time. This is due to an additional phase locking criterion that arises from having a total of three modes in the teleportation scheme as opposed to the two modes that are required for dense coding. This illustrates subtle intricacies of teleportation with mode entanglement of a massive particle that are not present in the particle entanglement case.

We begin in the subsequent section, by reviewing the concept of mode entanglement. In section (3), we detail how to teleport the unknown state of a spatial mode using a single massive particle distributed coherently across two spatial modes. We begin next by introducing the concept of spatial mode entanglement in more detail.

2   Entanglement of Spatial Modes

We will start by detailing the mode structure of a bosonic system and then use a simple example of mode entanglement to illustrate the differences between it and particle entanglement. Finally, we will explain some results concerning mode entanglement of Bose gases at zero and finite temperatures. Note that we will consider in this paper only mode entanglement of bosonic fields; for a discussion of mode entanglement in fermionic systems see, for example, this paper [39] by Aharonov and Vaidman.

Consider a confining volume, V, whose energy eigenmodes, labelled by k, can be 'excited' by applying creation operators, $\hat{a}_k^\dagger$, on the vacuum state, $\hat{a}_k^\dagger |vac\rangle = |0_1 0_2...1_k 0_{k+1}...\rangle$, where $[\hat{a}_k, \hat{a}_l^\dagger] = \delta_{kl}$. An excitation is 'a particle' with corresponding energy $E_k = \hbar \omega_k$, where ω_k is the frequency of the k-th energy eigenmode. As mentioned already, instead of describing the system using its energy modes, it is often desirable to use a different set of modes, such as the *spatial modes*. The description of the system in space is obtained by a transformation via the energy eigenfunctions, $\hat{a}_k^\dagger |vac\rangle = \int dx\, \phi_k(x) \hat{\psi}^\dagger(x) |vac\rangle$, where $\phi_k(x)$ is the k-th energy eigenfunction and $\hat{\psi}^\dagger(x)$ creates a particle at point, x, in space. It follows that $[\hat{\psi}(x), \hat{\psi}(x')] = \delta(x - x')$. Populating an energy mode with a particle is thus equivalent to populating all spatial modes, i.e. all points in space, in a superposed manner.

To illustrate the differences between particle and mode entanglement, we take two non-interacting bosons trapped in a confining volume at zero temperature. In first quantisation, i.e. in the language that one would use to describe particle entanglement, the wavefunction is the symmetrized product, $\Psi_{12}(x, y) = \frac{1}{\sqrt{2}}(\phi_1(x)\phi_2(y) + \phi_1(y)\phi_2(x))$, where $\phi(x)$ is the ground state of the confining potential. No entanglement exists between the particles, since indistinguishability forbids us from assigning to any particle a specific set of degrees of freedom. In other words, the state space of the two particles is a projection onto the symmetric subspace, whereas a tensor product Hilbert space, $\mathcal{H} = \mathcal{H}_1 \otimes \mathcal{H}_2$, is required to define entanglement between the subsystems.

Conversely, in second quantisation one can define a pair of spatial modes, A and B, where each mode occupies half the confining geometry. Since both the particles are *coherently* distributed over these modes, the system is described by the entangled state,

$$\begin{aligned}
|\psi\rangle &= \frac{(\hat{a}_0^\dagger)^2}{2}|vac\rangle \\
&= \frac{1}{2}\left(\frac{\hat{\psi}_A^\dagger + \hat{\psi}_B^\dagger}{\sqrt{2}}\right)^2 |vac\rangle \\
&= \frac{1}{2}(|20\rangle + \sqrt{2}|11\rangle + |02\rangle),
\end{aligned} \tag{1}$$

where $\hat{\psi}_X^\dagger = \int_X dx\, g(x)\hat{\psi}^\dagger(x)$ creates a particle in mode $X = A, B$ ($g(x)$ is a so called detector profile that gives weighting to the points in space), $|mn\rangle =$

$|m\rangle_A \otimes |n\rangle_B$ span the state space $\mathcal{H} = \mathcal{H}_A \otimes \mathcal{H}_B$ and m denotes the number of particles in mode A and n the number particles in mode B (with $m + n = 2$). From this example it is clear that entanglement is contingent on the choice of modes [41], but provided that like here a suitable choice is made, investigating entanglement between *distinguishable* modes circumvents the difficulties of defining entanglement between indistinguishable particles [40].

Spatial mode entanglement of an non-interacting BEC at zero temperature was first considered by Simon [27]. He and others [42] found that mode entanglement existed between regions of space if the gas had an uncertainty in particle number below a given level. That is, if the gas is best described by a coherent state, $|\alpha\rangle$ (or mixtures there of), there is no entanglement between spatial modes. On the other hand, if a uniform gas is of a fixed particle number, N, there is $\frac{1}{2}\log_2 N$ amount of entanglement (as measured by the von Neumann entropy), between two equal sized modes. Anders *et al* [28] used a thermodynamical entanglement witness to show that spatial mode entanglement only exists across an entire Bose gas below the critical temperature for BEC. Spatial entanglement between two and more modes of an interacting Bose gas at finite-temperature with a fixed (but possibly unknown) number of particles was considered by Goold *et al.* [31], who demonstrated that mode entanglement is present in a gas when the coherence length of the particles extends over the modes. More specifically, there is a direct link between single-particle reduced density matrix [43] (i.e. the visibility of interference fringes) between different regions of the gas and spatial mode entanglement. This means that the natural mode entanglement of a BEC has already been detected in experiments such as [44], albeit indirectly. More recently, mode entanglement generated by a single exciton was predicted to exist between different sites in the photosynthetic FMO complex [45], (as a result of the experimentally verified quantum coherence in the molecule), demonstrating that this type of entanglement is relevant even in biological systems.

3 Teleportation of a Quantum State of a Spatial Mode Using a Single Massive Particle

Since mode entanglement exists naturally within many systems, it is important to ask whether, at least in principle, this type of entanglement is useful entanglement. We will address the question by providing a scheme to show that a single massive particle that is coherently distributed over two spatial modes can, at least theoretically, be used to teleport the unknown quantum state of a spatial mode perfectly half of the time in spite of the superselection rule.

In the following, we will first introduce a Hamiltonian whose parameters can be switched on and off to perform the gates. We will show that in order to overcome the superselection rule and to rotate the modes away from the particle number basis, a Bose-Einstein condensate should be used as a particle reservoir and also as a phase reference throughout the teleportation protocol. We will end by introducing the quantum circuit that allows to perform the teleportation protocol.

The standard qubit teleportation protocol [34] between two parties, A for Alice and B for Bob, can be split up into four parts:

(i) **preparation stage:** the preparation of an unknown quantum state by a third party, Charlie, and also the distribution of an entangled Bell state between Alice and Bob,

(ii) **Bell state measurement by Alice:** the prepared qubit is passed to Alice who then makes a Bell state measurement on this and her portion of the entangled state and records the measurement outcome,

(iii) **transmission of classical information:** Alice sends Bob two bits of classical information that indicate which of the four Bell states was measured, and

(iv) **single mode rotation by Bob:** Bob uses the classical information to select which operation to perform on his qubit leaving him with the original unknown quantum state.

We will now detail how to perform the four steps using mode entanglement of a single massive particle.

The system consists of three spatial modes, a, A and B. Mode a will be placed in an unknown state, which will be teleported to mode B. Modes A and B will be in the maximally entangled state formed from a single particle. The system is described by the Bose-Hubbard model with additional coupling of each mode to the BEC reservoir. The Hamiltonian is written as follows

$$\hat{H}_{BH} = -\frac{J_{AB}}{2}(\hat{\psi}_A^\dagger \hat{\psi}_B + \hat{\psi}_B^\dagger \hat{\psi}_A) - \frac{J_{aA}}{2}(\hat{\psi}_a^\dagger \hat{\psi}_A + \hat{\psi}_A^\dagger \hat{\psi}_a)$$

$$+ \sum_{i=a,A,B} U_i \hat{n}_i(\hat{n}_i - 1) + \sum_{i=a,A,B} E_i \hat{n}_i - \sum_{i=a,A,B} \frac{\Omega_i}{2}(\hat{\psi}_i^\dagger \hat{\psi}_{res} + \hat{\psi}_{res}^\dagger \hat{\psi}_i) \quad (2)$$

The first two terms represent the coupling between the modes, A and B, and also modes, a and A. Modes a and B do not interact throughout the entire protocol. The coupling between the modes can be turned on and off by varying the tunneling matrix element J_{aA} (J_{AB}) by increasing/decreasing the height of the potential barriers between the wells by altering the intensity of the trapping laser.

The parameter U_i is the onsite interaction term, which is continuously set to be much larger than the other energy scale in the Hamiltonian. This is achieved by ensuring that each potential is tightly confined in all three directions, as is the case for atomic quantum dots [21]. The resulting large nonlinear repulsive interaction between the particles means that only zero or one particles can exist in the mode at any instance. In other words, by maintaining a high repulsivity between the particles the modes are forced to behave like qubits.

The next three terms in the Hamiltonian are the free energies E_i of the individual modes. The standard setting throughout the protocol will be $E_a = E_A = E_B$, but to change the phases of the modes relative to one another, a potential bias can be applied to a mode using, for instance, a dispersive laser pulse.

The final set of terms in the Hamiltonian correspond to the coupling of each individual mode to the BEC reservoir. We consider a Raman laser set up, which couples the two different trapping states of the system and reservoir atoms [46]. The parameter, $\Omega = \int dx \phi_0(x) \Psi_0(x) \tilde{\Omega}$, is the effective Rabi frequency, where $\Psi_0(x)$ is the wavefunction of the BEC, $\phi_0(x)$ is the wavefunction of the atom in mode a and $\tilde{\Omega}$ is the usual Rabi frequency.

3.1   Single-Mode and Two-Mode Gates

We will now discuss the implementation of the single-mode and two-mode gates that will be used in the teleportation protocol.

Single-mode phase gate: A phase gate, for instance a Pauli $\hat{Z}$ operation, can be applied to mode j by altering the energy, E_j, relative to the free energies for the other modes for a given time, t. The corresponding unitary operation on the mode is $\hat{U} = e^{iE_j \hat{n}_j t}$, which if applied for $t = \pi/E_i$ performs the $\hat{Z}$ gate, $|0\rangle \rightarrow |0\rangle$, $|1\rangle \rightarrow -|1\rangle$. During this operation, the couplings between the modes themselves and between the modes and the BEC are switched off.

Single-mode number rotation gate: We would also like to rotate the state of an individual mode between the $|0\rangle$ and $|1\rangle$ eigenstates. To do this we need to couple to the BEC reservoir by applying the Raman pulse for a time, t. We take here a uniform non-interacting BEC described by a mixture of coherent states, $\hat{\rho}_{bec} = \int_0^{2\pi} \frac{d\theta}{2\pi} ||\alpha|e^{i\theta}\rangle\langle|\alpha|e^{i\theta}|$, where $|\alpha|^2 = \bar{n}$ is the average number of particles in the condensate and θ is the condensate phase. For a nummerical analysis of a single mode, i.e. an atomic quantum dot, coupled to a superfluid reservoir of interacting particles see [47].

Note that when calculating it suffices to use the coherent state, $||\alpha|e^{i\theta}\rangle$, with one realisation of the phase θ and to apply the twirling operator, $T[\rho_L] = \int_0^{2\pi} \frac{d\theta}{2\pi}[\hat{\rho}_L]$, to obtain the quantum state of laboratory as seen by Bob (here $\hat{\rho}_L$ is the density operator for the total system and laboratory [20,17]). Note that, unlike the dense coding scheme [32], Alice and Bob do not necessarily need to use the same BEC for their operations in this teleportation scheme. The unitary evolution governed by $\hat{H}_{int} = \frac{\Omega_i}{2}(\hat{\psi}_i^\dagger \hat{\psi}_{res} + \hat{\psi}_{res}^\dagger \hat{\psi}_i)$, transforms the occupation number of the modes as follows

$$|0\rangle \rightarrow \cos(\frac{\Omega\sqrt{\bar{n}}t}{2})|0\rangle - i\sin(\frac{\Omega\sqrt{\bar{n}}t}{2})e^{i\theta}|1\rangle$$

$$|1\rangle \rightarrow \cos(\frac{\Omega\sqrt{\bar{n}}t}{2})|1\rangle - i\sin(\frac{\Omega\sqrt{\bar{n}}t}{2})e^{-i\theta}|0\rangle. \tag{3}$$

Here we have traced out the BEC, which remains separable to the state of the modes as we assume it to have a high mean number of particles, $\bar{n} \gg 1$. By coupling the modes to the BEC for various times, t, single mode number rotation gates can be performed.

Two-mode c-phase type gate: We also need a two mode entangling gate to allow for Bell state analysis. A c-phase type gate between two of the modes,

$j \neq k \in a, A, B$, is performed as follows. Since we consider a large onsite inter-action strength throughout this protocol, the bosons behave like spin-polarised fermions. One can see this by associating the two particle number states, $|0\rangle$ and $|1\rangle$, of each mode with the up/down spin half degree of freedom. The bosonic Hamiltonian, $\hat{H} = -\frac{J}{2}(\hat{\psi}_j^\dagger \hat{\psi}_k + \hat{\psi}_k^\dagger \hat{\psi}_j)$, becomes equivalent to the quantum XX spin model, which in turn can be mapped via the Jordan-Wigner transforma-tion to $\hat{H} = -J(\hat{c}_j^\dagger \hat{c}_k + \hat{c}_k^\dagger \hat{c}_j)$, where $\hat{c}_X^\dagger$ and $\hat{c}_X$ are Fermionic creation and annihilation operators for mode, $X = j, k$ that anti-commute, $\{\hat{c}_X, \hat{c}_Y\}_+ = \delta_{XY}$ [49]. Once the system is in this regime, the barrier between the modes, j and k, is lowered for time, $t = \pi/(2J)$, so that the particles exchange position. This ensures that the $|11\rangle$ term picks up a minus sign $|11\rangle \rightarrow -|11\rangle$, due to the anti-commutation relations [50]. The other three states transform as, $|00\rangle \rightarrow |00\rangle$ $|10\rangle \rightarrow |01\rangle$ and $|01\rangle \rightarrow |10\rangle$.

3.2 Circuit Diagram for Teleportation of a Quantum State of a Mode

We now present the circuit diagram for the teleportation protocol, see Fig. (1), referring to the four steps outlined at the start of this section. The three modes and the BEC are represented in the figure by the four horizontal lines in the diagram and the gates are represented as boxes.

The preparation stage: a third party, Charlie, prepares a spatial mode (a) in an unknown quantum state by performing a single-mode number rotation (green in the diagram [section (3.1)]) and a single-mode phase gate (red in the diagram [section (3.1)]) for the chosen amounts of time. Initially, mode a, is in the vacuum state $|a\rangle = |0\rangle_a$, so that the two gates rotate it to the state $|a\rangle = \alpha|0\rangle + \beta e^{i(\theta+\phi)}|1\rangle$, where $\alpha = \cos(\theta')$, $\beta = -i\sin(\theta')$ and $\theta' = \Omega\sqrt{\bar{n}}t/2$ and

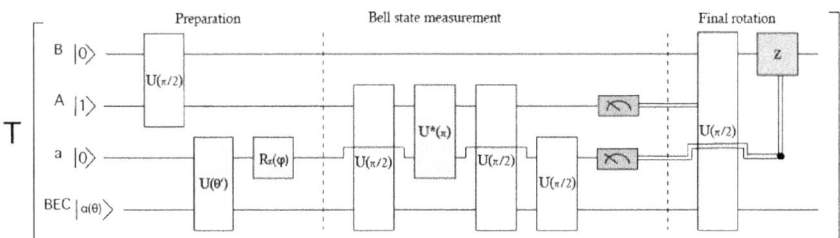

Fig. 1. The circuit for teleportation of a quantum state of a spatial mode. The three modes and the BEC are represented in the figure by the four horizontal lines in the diagram and the gates are represented as boxes. The first portion of the circuit is the *preparation* of the entanglement via a single particle between Alice and Bob and the creation of an unknown state of the spatial mode a. The *Bell state measurement* on modes a and A by Alice is outlined in the centre portion of the circuit. The last part of the circuit details the *final rotation* on mode B by Bob. Further details will be discussed in the text.

$\phi = E_i t$. The potential barrier between modes a and A remains high throughout this stage. An entangled Bell state shared between Alice and Bob is also required for the protocol. To obtain such a state a single massive particle needs to be coherently distributed over the two modes – if the single neutral atom initially starts in mode A, lowering the barrier to mode B for time $t = \pi/(2J)$ creates the state $|\psi^+\rangle_{AB} = \frac{1}{\sqrt{2}}(|10\rangle + |01\rangle)$.

The Bell state measurement: Spatial mode a is now passed to Alice, who performs Bell state analysis on modes, a and A. Bob is in possession of mode B. A measurement in the Bell state basis proceeds as follows:

(1) Modes a and A should be within close proximity of one another and the potential barrier between them is high so that tunneling is fully suppressed. Mode A is rotated away from the particle number basis by coupling to a BEC reservoir via the Raman laser set-up for time $t = \pi/(2\Omega\sqrt{\bar{n}})$ as denoted by the yellow gate in the circuit diagram [see section (3.1)]. Note that the BEC here does not necessarily need to be the same as the one used by Charlie to create the initial state of mode a. This results in the transformations, $|0\rangle_A \to \frac{1}{\sqrt{2}}(|0\rangle - ie^{i\theta}|1\rangle)_A$ and $|1\rangle_A \to \frac{1}{\sqrt{2}}(|1\rangle - ie^{-i\theta}|0\rangle)_A$. For instance, the $|\psi_+\rangle_{aA}$ Bell state transforms (without normalisation) to $|01\rangle_{aA} + |10\rangle_{aA} \to |0\rangle_a(|1\rangle - ie^{-i\theta}|0\rangle)_A + |1\rangle_a(|0\rangle - ie^{i\theta}|1\rangle)_A$. The other Bell states transform in a similar way.

(2) Then, a c-phase type gate [see section (3.1)] is applied by lowering the potential barrier between modes a and A for time $t = \pi/(2J)$ so that the bosons swap positions. This is illustrated by the pale blue gate in the circuit diagram. By continuing the example in the previous step, one can see that this transformation works in the following way: $|0\rangle_a(|1\rangle - ie^{-i\theta}|0\rangle)_A + |1\rangle_a(|0\rangle - ie^{i\theta}|1\rangle)_A \to |01\rangle_{aA} + |10\rangle_{aA} - ie^{-i\theta}|00\rangle_{aA} + ie^{i\theta}|11\rangle_{aA} = (|0\rangle + ie^{i\theta}|1\rangle)_a(|1\rangle - ie^{-i\theta}|0\rangle)_A$.

(3) In order to rotate modes, a and A, to the particle number basis for the read-out, each should be coupled to the BEC reservoir for time, $t = \pi/(2\Omega\sqrt{\bar{n}})$, as in step (2). For example, the state, $(|0\rangle + ie^{i\theta}|1\rangle)_a(|1\rangle - ie^{-i\theta}|0\rangle)_A$, after c-phase type gate above becomes $|00\rangle_{aA}$. Hence a measurement of zero particles in both modes means that the quantum state, $|a\rangle = \alpha|0\rangle + \beta e^{i(\theta+\phi)}|1\rangle$ has been teleported to Bob. A measurement of $|01\rangle_{aA}$ corresponds to the state, $\hat{Z}|a\rangle = \alpha|0\rangle - \beta e^{i(\theta+\phi)}|1\rangle$, being teleported to Bob's mode. Note that the teleported states both also have phases which are correlated to the BEC, which means if they are to be used for other purposes then the *same* BEC would be needed to ensure phase matched conditions.

However, if there is one particle in mode a after this final rotation, the teleportation has not succeeded. This is due to the fact that after steps (1)-(3), the states, $|\phi^\pm\rangle_{aA}$, are rotated to $|1\rangle_a((1\pm e^{i2\theta})|0\rangle + (1\mp e^{i2\theta})|1\rangle)_A/2$. Since one does not know the phase, θ of the condensate, one has to average over all realisations of it, i.e. apply the twirling operator to determine the quantum state of the laboratory, which accounts for our ignorance. This leaves mode A in the maximally mixed state so that it is impossible to say which of the two operators, $\hat{X}$ and $\hat{Z}\hat{X}$, should be applied to recover the original state, $|a\rangle$, even in principle.

Transmission of classical information and final single-mode rotation:
Alice thus sends one of three messages to Bob according to the outcome of her
measurement. Bob obtains the unknown quantum state half of the time upon
the following transformations:

$$|\psi^+\rangle_{aA} \rightarrow |00\rangle_{aA} \rightarrow \hat{I}_B, \qquad |\psi^-\rangle_{aA} \rightarrow |01\rangle_{aA} \rightarrow \hat{Z}_B$$
$$|\phi^\pm\rangle_{aA} \rightarrow |1\rangle\langle 1|_a \otimes \hat{I}_A \quad \rightarrow \quad \text{teleportation failed} \tag{4}$$

where the first column corresponds to the Bell state Alice measured, the second
column to the particle number measurements in mode a and A (and also the
classical bits that she sends) and the third column to Bob's operation on mode
B. The $\hat{Z}$ operation is applied to mode B by changing the phase of the mode [as
in section (3.1)] for time $t = \pi/E_B$ where E_B is the energy bias of mode B. This
concludes the teleportation protocol for a unknown quantum state of a spatial
mode.

4   Discussion and Conclusion

In this paper, we have discussed the teleportation of an unknown state of a
spatial mode using a mode entangled state formed from a single particle. We
modeled the modes as qubits by considering tightly confined potentials, with
zero or one particles representing the qubit degrees of freedom. We have given
an explicit scheme for creating an arbitrary state of a spatial mode despite the
particle number superselection rule, by coupling to a BEC reservoir. We have
shown that one can reliably distinguish two of the four Bell states, with the other
two grouped together. This is due to the fact that we do not know the phase
of the BEC reservoir, which is still classically correlated to the state after the
Bell state analysis. However, the teleportation of a general unknown (two-level)
state of a spatial mode is still achieved half of the time, which would never be
possible if we could not locally bypass the superselection rule in the first place.

   This contrasts with a dense coding scheme using the mode entanglement of
a single particle (see [32]). In this scheme, one can always discriminate all four
Bell states. The difference between the two schemes arises from the difference in
symmetries of the encoding (or preparation) and decoding (or Bell state analysis)
processes in the two protocols. In the dense coding protocol, there are only two
modes and Alice can initially couple her mode to the BEC to rotate to one of
the two states, $|\phi^\pm\rangle = \frac{1}{\sqrt{2}}(|00\rangle \pm e^{i2\theta}|11\rangle)$. In doing so, the phase of these states
becomes classically correlated to the phase of the BEC, θ. Alice then passes her
mode to Bob, who performs Bell state analysis as was outlined above to find
which Bell state Alice sent. Because the phase of the BEC is already present
in $|\phi^\pm\rangle$ it cancels with the phase that is picked up again from the BEC during
the Bell state measurement. Thus all four Bell states are reliably distinguishable
from one another.

   On the other hand, in the teleportation scheme, there are necessarily three
modes. Initially the mode that will be subsequently teleported to Bob is rotated

to an arbitrary state and becomes classically correlated to the phase of the BEC – *this phase will stay present in the state of this mode as it is teleported to Bob and does not feature in the Bell state analysis.* Bell state analysis is then performed on this and one mode of the entangled state. However, the phases that are picked up from the BEC in the Bell state analysis do not have any phases to cancel with, unlike in the case of the dense coding scheme. So while we have access to all of the gates that are required to perform the teleportation protocol, because we cannot know the phase of the BEC, even in principle, we do not teleport the spatial mode all of the time. This paper has illustrated how mode entanglement of massive particles does not behave exactly like particle entanglement despite the fact that one can locally overcome the particle number superselection rule. Future work should focus on extending this scheme to consider the teleportation of a general $d > 2$ level state of a spatial mode using the natural entanglement in BECs.

References

1. Nielsen, M., Chuang, I.: Quantum information and quantum computation (2000)
2. Zanardi, P.: Phys. Rev. A 65, 042101 (2001)
3. Tichy, M.C., de Melo, F., Kus, M., Mintert, F., Buchleitner, A., arXiv:0902.1684v5
4. Peres, A.: Phys. Rev. Lett. 74, 4571 (1995)
5. Tan, S.M., Walls, D.F., Collett, M.J.: Phys. Rev. Lett. 66, 252 (1991)
6. van Enk, S.: Phys. Rev. A 72, 064306 (2006)
7. Hardy, L.: Phys. Rev. Lett. 73, 2279 (1994)
8. Greenberger, D.M., Horne, M.A., Zeilinger, A.: Phys. Rev. Lett. 75, 2064 (1995)
9. Dunningham, J., Vedral, V.: Phys. Rev. Lett. 99, 180404 (2007)
10. Heaney, L., Cabello, A., Santos, M.F., Vedral, V., arXiv:0911.0770v2
11. Hessmo, B., Usachev, P., Heydari, H., Björk, G.: Phys. Rev. Lett. 92, 180401 (2004)
12. Papp, S.B., Choi, K.S., Deng, H., Lougovski, P., van Enk, S.J., Kimble, H.J.: Science 324, 764 (2009)
13. Wiseman, H.M., Vaccaro, J.A.: Phys. Rev. Lett. 91, 097902 (2003)
14. Wick, G.C., Wightman, A.S., Wigner, E.P.: Phys. Rev. 88, 101 (1952)
15. Giulini, D., Joos, E., Kiefer, C., Kupsch, J., Stamatescu, I.-O., Zeh, H.D.: Decoherence and the appearance of the classical world in quantum theory. Springer, Heidelberg (1996)
16. Greenberger, D.M., Horne, M.A., Zeilinger, A.: Quantum Interferometry. In: De-Martini, F., et al. (eds.). VCH Publishers (1996)
17. Bartlett, S.D., Rudolph, T., Spekkens, R.W.: Rev. Mod. Phys. 79, 555 (2007)
18. Aharonov, Y., Susskind, L.: Phys. Rev. 155, 1428 (1967)
19. Mirman, R.: Phys. Rev. 186, 1380 (1969)
20. Dowling, M.R., Bartlett, S.D., Rudolph, T., Spekkens, R.W.: Phys. Rev. A 74, 052113 (2006)
21. Recati, A., et al.: Phys. Rev. Lett. 94, 040404 (2005)
22. Terra Cunha, M.O., Dunningham, J.A., Vedral, V.: Proc. Royal Soc., A 463, 2277 (2007)
23. Ashhab, S., Maruyama, K., Nori, F.: Phys. Rev. A 75, 022108 (2007)
24. Heaney, L., Anders, J.: Phys. Rev. A 80, 032104 (2009)
25. Ashhab, S., Maruyama, K., Brukner, Č., Nori, F.: Phys. Rev. A 80, 062106 (2009)

26. Heaney, L., Lee, S.-.W., Jaksch, D.: Phys. Rev. A 82, 042116 (2010)
27. Simon, C.: Phys. Rev. A 66, 052323 (2002)
28. Anders, J., Kaszlikowski, D., Lunkes, C., Ohshima, T., Vedral, V.: N. J. Phys. 8, 140 (2006)
29. Heaney, L., Anders, J., Kaszlikowski, D., Vedral, V.: Phys. Rev. A 76, 053605 (2007)
30. Heaney, L.: PhD thesis, University of Leeds (2008)
31. Goold, J., Heaney, L., Busch, T., Vedral, V.: Phys. Rev. A 80, 22338 (2009)
32. Heaney, L., Vedral, V.: Phys. Rev. Lett. 103, 200502 (2009)
33. Bennett, C.H., Wiesner, S.J.: Phys. Rev. Lett. 69, 2881 (1992)
34. Bennett, C.H., Brassard, G., Crepeau, C., Jozsa, R., Peres, A., Wootters, W.: Phys. Rev. Lett. 70, 1895 (1993)
35. Bouwmeester, D., Pan, J.M., Mattle, K., Eible, M., Weinfurter, H., Zeilinger, A.: Nature (London) 390, 575 (1997)
36. Furusawa, A., Sorensen, J.L., Braunstein, S.L., Fuchs, C.A., Kimble, H.J., Polzik, E.S.: Science 282, 706 (1998)
37. Boschi, D., Branca, S., De Martini, F., Hardy, L., Popescu, S.: Phys. Rev. Lett. 80, 1121 (1998)
38. Lombardi, E., Sciarrino, F., Popescu, S., De Martini, F.: Phys. Rev. Lett. 88, 070402 (2002)
39. Aharonov, Y., Vaidman, L.: Phys. Rev. A 61, 052108 (2000)
40. Estéve, J., Gross, C., Weller, A., Giovanazzi, S., Oberthaler, M.K.: Nature 455, 1216 (2008)
41. Vedral, V.: Cent. Euro. J. Phys. 2, 289 (2003)
42. Tóth, G., Simon, C., Cirac, J.I.: Phys. Rev. A 68, 062310 (2003)
43. Pitaevskii, L., Stringari, G.: Bose-Einstein condensation. Springer, Heidelberg (2003)
44. Bloch, I., Hänsch, T.W., Esslinger, T.: Nature 403, 166 (2000)
45. Sarovar, M., Ishizaki, A., Fleming, G.R., Whaley, K.B.: Nature Physics 6, 462 (2010)
46. Jaksch, D., Zoller, P.: Annals of Physics 315, 52 (2005)
47. Lee, H.-J., Byczuk, K., Bulla, R.: Phys. Rev. B 82, 054516 (2010)
48. Note that controlling the phase of a BEC in this way is known as phase imprinting; see e.g. [51]
49. Sachdev, S.: Quantum phase transitions. Cambridge University Press, Cambridge (2000)
50. Clark, S., Moura Alves, C., Jaksch, D.: N. J. Phys. 7, 124 (2005)
51. Denschlag, J., et al.: Science 287, 97 (2000)

Author Index